Smart Supply Chain Management

Muhammad Shujaat Mubarik ·
Sharfuddin Ahmed Khan
Editors

Smart Supply Chain Management

Design, Methods and Impacts

Springer

Editors
Muhammad Shujaat Mubarik
Edinburgh Business School
Heriot-Watt University
Edinburgh, UK

Sharfuddin Ahmed Khan
University of Regina
Regina, SK, Canada

ISBN 978-981-96-1332-8 ISBN 978-981-96-1333-5 (eBook)
https://doi.org/10.1007/978-981-96-1333-5

This Springer imprint is published by the registered company Springer Nature Singapore Pte Ltd.
The registered company address is: 152 Beach Road, #21-01/04 Gateway East, Singapore 189721, Singapore

Acknowledgements

We are thankful to the Higher Education Commission (HEC) of Pakistan, NRPU: 20-11226, Institute of Business Management (IoBM) Karachi, Edinburgh Business School, Heriott-Watt University, UK, and University of Regina, Canada, for their support.

Introduction

The complexity of modern supply chains, characterized by their global reach and interconnectedness, has made them increasingly vulnerable to disruptions. Natural disasters, geopolitical tensions, pandemics, and fluctuating market demands have exposed the fragility of traditional supply chain models, causing significant delays, increased costs, and operational inefficiencies. The COVID-19 pandemic, for instance, disrupted supply chains worldwide, highlighting the need for more resilient and adaptive systems. In this context, the concept of the smart supply chain has emerged as a promising solution, leveraging advanced digital technologies to create more efficient, responsive, and sustainable supply chains. The impetus for editing this book stems from the urgent need to address these challenges and explore how smart supply chains can transform traditional supply chain management. The book seeks to provide a comprehensive understanding of how integrating digital technologies such as the Internet of Things (IoT), artificial intelligence (AI), blockchain, and big data analytics can revolutionize supply chains, making them more resilient and capable of responding to the complexities of the modern business environment. The rapid pace of technological advancements and the increasing pressure on companies to enhance their supply chain performance have made it essential to rethink supply chain management from a smarter, more innovative perspective. This book aims to fill that gap by offering insights into the theory, methods, and practical applications of smart supply chains. The book begins by examining the current state of supply chains, highlighting the numerous challenges they face in today's interconnected world. These challenges are not just operational but also strategic, as companies must navigate a global landscape filled with uncertainties and disruptions. The discussion sets the stage by emphasizing that while globalization has enabled supply chains to operate on a larger scale, it has also introduced vulnerabilities that can have cascading effects across industries. For example, a disruption in a single supplier's operations can lead to significant delays and shortages across the entire supply chain, as was evident during the semiconductor shortage that impacted the automotive and electronics industries worldwide.

To address these issues, the concept of the smart supply chain is introduced as a solution that leverages cutting-edge technologies to enhance visibility, predictability, and efficiency. Smart supply chains are not merely an incremental improvement over traditional models; they represent a fundamental shift in how supply chains are designed and managed. By incorporating IoT, AI, blockchain, and other advanced technologies, smart supply chains can provide real-time data, automate decision-making processes, and optimize operations, thereby significantly reducing the risks associated with disruptions. The book delves into how these technologies can be integrated into supply chain operations to create systems that are more responsive, resilient, and capable of adapting to changes in the environment. One of the key focuses of the book is the role of AI in smart supply chain management. AI's ability to process large volumes of data and generate predictive insights is explored in depth, showcasing how it can enhance demand forecasting, inventory management, and supply chain visibility. The discussion highlights that while AI offers immense potential, its adoption comes with challenges, such as the need for skilled personnel and concerns around data security and ethical considerations. The book addresses these barriers and provides strategies for overcoming them, making AI a practical tool for enhancing supply chain performance.

In addition to AI, the book explores the impact of other emerging disruptive technologies on supply chain management. Blockchain and IoT are discussed as critical enablers of smart supply chains. Blockchain's role in providing transparency and traceability is particularly emphasized, as it ensures that every transaction within the supply chain is secure and verifiable. This capability is crucial for industries that require high levels of trust and compliance, such as pharmaceuticals and food safety. IoT's contribution to real-time monitoring and data collection is also explored, illustrating how it enhances supply chain visibility and allows for proactive management of potential issues. The book further discusses how smart supply chains can capitalize on innovation capabilities through collaborative innovation networks. This section underscores the importance of collaboration among various stakeholders, including suppliers, manufacturers, and customers, in driving continuous improvement and fostering innovation within the supply chain. The discussion also emphasizes the need for a robust knowledge management framework to support the effective implementation of smart supply chain technologies. As the book progresses, it shifts focus to specific strategies and applications within smart supply chain management. The use of queuing strategies in smart supply chains is examined, highlighting how these strategies can optimize workflows and reduce bottlenecks, thereby improving overall efficiency. Ethical considerations are also addressed, with a dedicated discussion on how smart supply chain management can promote ethical practices across the supply chain, ensuring compliance with regulations and upholding social and environmental responsibilities. Leadership in the context of smart supply chain management is another critical area explored in the book. The complexity of integrating advanced technologies into supply chain operations requires effective leadership to navigate the associated challenges. The book provides insights into contemporary leadership practices that align with the goals of smart supply chain management, offering guidance on how to lead organizations through this transformative process. The role of knowledge management in smart supply chains is also given significant attention. The book argues that effective knowledge management is crucial for the successful implementation of smart supply chain technologies. By facilitating the seamless sharing of information and enhancing decision-making processes, knowledge management helps organizations optimize their supply chain operations and achieve sustainable competitive advantages. Toward the end, the book looks forward, providing a roadmap for the future of smart supply chains. It discusses emerging technologies and trends that are likely to shape the future of supply chain management, encouraging readers to consider the long-term implications of adopting smart technologies. The book emphasizes the importance of staying ahead of technological advancements to maintain a competitive edge in the global market.

In the end, let us emphasize that smart technologies will shape the future of supply chains in the form of overproduction reduction, shipment consolidation, and waste reduction. Moreover, the supply chains will be well-connected and traceable. The actors in the supply chain—*producer, intermediaries, processor, logistics, distributor, retailer, and customer*—will be connected through IoT and blockchain. Besides connectivity, the IoT and blockchain will adapt to consumer requirements. In the future, supply chain actors will have unrestricted access to IoT networks and distributed clouds. Because of being environmentally sensitive, digitalized, resource-efficient, traceable, and well-connected, the futuristic supply chain functions must be as resilient as possible. Finally, since developed economies and third-world countries are not parallel in technology adoption, policymakers must design localized strategies.

Contents

About the Editors

Muhammad Shujaat Mubarik is Associate Professor of Logistics and Supply Chain Management at Edinburgh Business School, Heriot-Watt University, UK. He is also a Visiting Scholar at the Department of Land Economy, University of Cambridge, UK, and has previously served as a Visiting Professor at Copenhagen Business School, Denmark. With an extensive academic and leadership background, Dr. Mubarik has held key positions, including Professor and Dean of the College of Business Management at the Institute of Business Management (IoBM), Karachi. He also serves as the President of the Karachi Chapter of The Association of Management Development Institutions in Pakistan (AMDIP). His research interests span both behavioural and technological dimensions of supply chains, with a strong focus on intellectual capital, sustainability, and digital transformation. His work explores the role of supply chain mapping and digital twin technologies in enhancing the triple bottom line (economic, environmental, and social sustainability) of supply chains. Dr. Mubarik's research has been published in leading journals, including Supply Chain Management: An International Journal, International Journal of Production Research, Technological Forecasting and Social Change, and Business Strategy and the Environment, among others. He has authored three books and edited two books and has served as a Principal Investigator (PI) and Co-PI on multiple research projects funded by various agencies. Beyond academia, Dr. Mubarik is a frequent media commentator, regularly appearing on television to provide expert insights on national and international economic and business affairs.

Sharfuddin Ahmed Khan is Assistant Professor in Industrial and Systems Engineering at Faculty of Engineering and Applied Science, University of Regina, SK, Canada. Prior to joining the University of Regina, Dr. Khan has worked more than twelve years with Industrial Engineering and Engineering Management department at the University of Sharjah (UoS), Sharjah, United Arab Emirates, where he has served as Assistant Professor (September 2019 to December 2021) and Lecturer (September 2009 to August 2019). Dr. Khan has more than 100 archival publications to his credit in high impact factor journals, as well as conference papers/ proceedings, and books chapters. He's published in various journals such as *Supply*

Chain Management: An International Journal, *International Journal of Production Research*, *IEEE Transactions on Engineering Management*, *Journal of Cleaner Production*, *Technological Forecasting and Social Change*, *Production Planning and Control*, *Experts Systems with Application*, *Journal of Operation Research Society*, *Annals of Operations Research*, *International Journal of Information Technology and Decision Making*, *Corporate Social Responsibility and Environmental Management*, and *Knowledge-Based Systems,* etc. He also attended several international conferences such as POMS, IEEE-IEEM, IEOM, GOL, etc. Sharfuddin has also guest edited special issues of prestigious international journals in the areas of Supply Chain Management and Decision Making.

Chapter 1
Embracing the Smart Revolution: Smart Supply Chain Management

Muhammad Shujaat Mubarik and Sharfuddin Ahmed Khan

1.1 Introduction

The contemporary global SC landscape is beset by an array of formidable challenges, each one capable of significantly disrupting the flow of goods and services across the world. Natural calamities such as earthquakes, tsunamis, and hurricanes can devastate infrastructure, halt production, and sever supply routes. For example, the 2011 earthquake and tsunami in Japan wreaked havoc on global SCs, particularly in the automotive and electronics industries, as the country is a critical supplier of components for these sectors. Similarly, international conflicts, trade wars, and geopolitical tensions have become increasingly common disruptors (Mubarik and Khan 2024). The USA-China trade war, with its imposition of tariffs on billions of dollars' worth of goods, forced companies across the globe to reconsider their SC strategies, leading to delays, increased costs, and even the rerouting of production lines (Fan et al. 2022; Agrawal et al. 2024). The COVID-19 pandemic further highlighted the fragility of modern SCs, causing unprecedented disruptions as lockdowns, border closures, and shifts in consumer demand brought many industries to a standstill (Mubarik et al. 2021; Ivanov and Dolugi 2021). These examples underscore the reality that today's SCs are deeply interconnected on a global scale. A single product might be composed of components sourced from multiple countries, each contributing to a final assembly that could take place thousands of miles away from where the product is eventually sold. This global nature of SCs, a byproduct of globalization, has made them more efficient but also more vulnerable to a variety of disruptions. A factory shutdown in one part of the world can ripple across continents, causing production delays

M. S. Mubarik (✉)
Edinburgh Business School, Heriot-Watt University, Edinburgh, UK
e-mail: shujaatmubarik@gmail.com

Institute of Business Management, Karachi, Pakistan

S. A. Khan
University of Regina, Regina, Canada

M. S. Mubarik and S. A. Khan (eds.), *Smart Supply Chain Management*,
https://doi.org/10.1007/978-981-96-1333-5_1

and shortages elsewhere. For instance, the 2021 semiconductor shortage, initially triggered by production halts in Asia, severely impacted the automotive industry worldwide, forcing automakers in the USA and Europe to slow down or temporarily cease production.

Within this complex and interwoven SC environment, three major challenges have emerged as particularly critical: *sustainability, resilience, and performance*. SC Sustainability has become a pressing concern as consumers and regulators increasingly demand environmentally and socially responsible practices. SCs must now consider not just the cost and speed of production but also the environmental impact and ethical sourcing of materials. Resilience of SC, or the ability to prepare for, respond to, and recover from disruptions, has also taken centre stage, especially considering recent global events that have exposed the weaknesses in traditional SC models. Finally, SC performance remains a constant pressure, as companies strive to meet consumer demands for faster, cheaper, and higher-quality products while navigating these other challenges.

In the face of these daunting issues, digital technologies (DTs) have emerged as a beacon of hope, offering new ways to tackle the complexities of modern SCM. These technologies, which we can refer to as *"smart technologies"*, present a significant opportunity to transform SCs into more agile, efficient, and resilient systems. By leveraging advancements in artificial intelligence (AI), the Internet of Things (IoT), big data analytics, blockchain, and automation, companies can gain unprecedented visibility into their SCs, optimize operations, and make data-driven decisions in real-time. For instance, IoT sensors can monitor the condition of goods in transit, ensuring that they are stored at the correct temperature and alerting stakeholders to any deviations that could affect product quality (Ivanov et al. 2019; Mubarik and Shahbaz 2023). AI-powered analytics can predict demand more accurately, helping companies to avoid stockouts or overproduction, while blockchain technology can enhance transparency and traceability, ensuring that every step in the SC meets ethical and regulatory standards.

These smart technologies are not just incremental improvements but potential disruptors that can redefine the very nature of SCs (Zhang et al. 2023). The incorporation of digital technologies into SC processes allows for the automation of repetitive tasks, reducing the risk of human error and freeing up human capital for more strategic roles. For example, robotic process automation (RPA) can handle routine tasks such as order processing and inventory management with greater accuracy and speed than human workers. Similarly, AI can be used to optimize routing and scheduling in logistics, minimizing delays and reducing transportation costs (Agrawal et al. 2024). The smart SC, therefore, represents a new paradigm in which digital technologies are deeply integrated into every aspect of SCM, from procurement and production to logistics and distribution. A smart SC can be defined as a digitally enhanced, data-driven, and highly responsive system that leverages advanced technologies to optimize every stage of the SC. It is characterized by real-time data flows, predictive analytics, automation, and a high degree of connectivity among all stakeholders. This transformation from traditional to smart SCs is not just about adopting new technologies; it is about rethinking the entire SC model to create a system that is more resilient, sustainable, and efficient. This new approach provides the impetus for the present book, which seeks to explore the theory, methods, and design of smart SCM.

As the first chapter of this book, the discussion here is intended to lay the groundwork for understanding the various facets of smart SCs. It aims to provide readers with key insights into what smart SCs are, why they are important, and how they can be leveraged to address the challenges faced by modern SCs. This chapter serves as a gateway to the more detailed discussions that will follow in subsequent chapters, offering an overview of the concepts, technologies, and methodologies that underpin smart SCM. By examining the global nature of SCs, the vulnerabilities they face, and the opportunities presented by digital technologies, this chapter sets the stage for a deeper dive into the specific strategies and tools that can be employed to build smart SCs. It also highlights the critical importance of sustainability, resilience, and performance in shaping the future of SCM.

The chapter starts with an examination of the challenges faced by contemporary SCs and the impact of globalization on SC vulnerabilities. It then introduces the concept of smart technologies as a solution to these challenges, providing examples of how these technologies can be applied in practice. The chapter concludes by defining the smart SC and explaining its significance in the context of modern SCM. This structure is designed to guide the reader from an understanding of the problems at hand to the exploration of potential solutions, ultimately providing a solid foundation for the rest of the book.

1.2 Supply Chains in the Present Era

In the present era, SCs are no longer seen merely as linear sequences of processes, but rather as complex, dynamic networks of organizations working in concert to fulfil the needs of consumers. These networks span across the globe, integrating various entities such as suppliers, manufacturers, distributors, retailers, and service providers. The interconnectedness of these organizations is facilitated by the flow of information, materials, and finances, creating a highly coordinated system that transcends geographical boundaries (Shi et al. 2023). This global integration allows companies to source raw materials from one continent, manufacture products in another, and distribute them to consumers worldwide, enabling businesses to operate efficiently on a global scale. The evolution of SCs into these intricate networks has made them the backbone of modern organizations. The economic progress seen in recent decades can be largely attributed to the efficiencies brought about by well-managed SCs. For instance, companies like Amazon, Tesla, P&G, Unilever, Walmart, and Apple to name a few, have revolutionized their respective industries by mastering their SCs. Amazon's ability to deliver products quickly and efficiently to consumers around the world is a direct result of its highly integrated and technologically advanced SC. Similarly, Apple's global SC, which involves suppliers and manufacturers from numerous countries, has been critical in allowing the company to consistently bring innovative products to market at a scale and speed that competitors struggle to match. These examples illustrate how modern SCs are central to the success and economic viability of contemporary businesses.

However, as SCs have become more integral to business operations, they have also become increasingly susceptible to a range of unprecedented challenges. The globalization that has enabled companies to optimize costs and efficiency has also introduced vulnerabilities that can have far-reaching consequences (Zhu et al. 2024). Disruptions in one part of the world can now quickly affect the entire SC, leading to delays, increased costs, and even halts in production. The COVID-19 pandemic is a stark reminder of this, where lockdowns and restrictions in key manufacturing hubs led to widespread shortages and disruptions across various industries, from electronics to automotive. Another significant challenge is the growing demand for sustainability. Consumers, governments, and other stakeholders are placing increasing pressure on companies to adopt environmentally friendly practices, reduce carbon footprints, and ensure ethical sourcing. Achieving these goals within a global SC is complex and often requires significant changes to traditional practices. Technological advancements have always been at the heart of SC evolution, with each era bringing new tools and methods to improve efficiency and meet emerging challenges. The adoption of technology in SCs is not a new phenomenon; whenever new technologies have emerged; they have been integrated into SC operations to enhance performance (Boudouaia et al. 2024). For example, the introduction of the steam engine and railways in the nineteenth century revolutionized transportation, enabling goods to be moved faster and more efficiently across long distances. In the twentieth century, the advent of computers and information technology allowed for more precise inventory management and data-driven decision-making, which significantly improved the responsiveness and efficiency of SCs (AlMulhim 2021; Mubarik and Khan 2024).

In the current era, we are witnessing the rise of modern digital technologies (DTs), particularly those associated with Industry 4.0 and Industry 5.0, which offer unprecedented computational, analytical, and robotic capabilities to organizations. Industry 4.0, characterized by the integration of cyber-physical systems, IoT, cloud computing, and big data analytics, is transforming SCs into highly automated and interconnected systems. These technologies allow for real-time monitoring and control of SC activities, predictive maintenance of machinery, and enhanced data analytics for decision-making. Industry 5.0 takes this a step further by emphasizing human–robot collaboration, where advanced robotics and AI work alongside human workers to enhance productivity and customization. These advancements are enabling organizations to create more responsive, flexible, and efficient SCs (AlMulhim 2021; Khan and Mubarik 2024).

The pace and scale of these unprecedented challenges have propelled organizations to adopt these modern DTs more rapidly than ever before. The disruptions caused by global events, the increasing demand for sustainability, and the need for greater resilience in SCs have all highlighted the limitations of traditional SC models. As a result, companies are turning to digital technologies to help them navigate these complex challenges. For instance, AI and machine learning are being used to improve demand forecasting, optimize inventory levels, and enhance SC visibility (AlMulhim 2021). Blockchain technology is being deployed to increase transparency and traceability, particularly in industries where ethical sourcing and compliance are critical. Robotics and automation are being implemented to improve efficiency and reduce reliance on human labour in certain areas of the SC. Nevertheless, the adoption of these DTs is not without its challenges. Implementing modern DTs in SCs requires an exhaustive, multidimensional analysis and evaluation. Companies must consider not only the technical feasibility of these technologies but also their impact on the overall SC, including the potential risks and benefits. For example, while automation can improve efficiency, it can also lead to job displacement and require significant investments in new infrastructure and training. Similarly, while blockchain can enhance transparency, it also raises concerns about data security and privacy (Zhang et al. 2022, Mubarik and Khan 2024). Therefore, organizations must carefully assess the potential impact of these technologies on their SCs and develop strategies to manage the associated risks. The emergence of the smart SC, driven by the integration of DTs, is still in its early stages, and its full potential is yet to be realized. Smart SCs leverage advanced technologies to create systems that are more intelligent, agile, and responsive to changes in the environment. However, implementing a smart SC is a complex process that requires careful planning and execution. Organizations must not only invest in the necessary technologies but also develop the skills and capabilities needed to manage and operate these systems effectively. This requires a comprehensive understanding of the underlying technologies, as well as the ability to integrate them into existing SC operations.

Given that the concept of the smart SC is relatively new, there is a need for exhaustive analysis and evaluation to ensure its successful implementation. This involves assessing the readiness of the organization to adopt these technologies, understanding the potential impact on the SC, and developing strategies to manage the associated risks. It also requires a thorough understanding of the different technologies involved, their potential applications, and the challenges associated with their implementation.

1.3 Smart Supply Chain: What It is?

Smart SCs are distinguished by their integration of advanced digital technologies that enable enhanced efficiency, agility, and responsiveness throughout the entire SC. These technologies, often referred to as "smart technologies", include the Internet of Things (IoT), AI, machine learning, blockchain, big data analytics, and advanced robotics. Each of these technologies plays a critical role in transforming traditional SCs into smart systems capable of real-time data processing, predictive analytics, and autonomous decision-making (Zhang et al. 2022). IoT, for example, allows physical objects within the SC—such as vehicles, machinery, and products—to communicate and share data through embedded sensors and connectivity. This constant flow of information provides unprecedented visibility into every stage of the SC, from raw material sourcing to final product delivery (Lee et al. 2023). With IoT, companies can monitor the condition of goods in transit, track inventory levels in real-time, and quickly respond to any disruptions or anomalies.

AI and machine learning further enhance the intelligence of the SC by analysing the vast amounts of data generated by IoT devices and other sources. These technologies enable predictive analytics, which can forecast demand, optimize inventory management, and improve production scheduling. By learning from historical data and continuously updating predictions, AI helps companies make more informed decisions, reduce waste, and increase overall efficiency. Blockchain technology adds another layer of innovation by providing a secure, decentralized ledger that records every transaction and movement within the SC. This transparency ensures that all parties have access to accurate and immutable records, reducing the risk of fraud, enhancing traceability, and improving compliance with regulatory requirements (Vahidinia et al. 2023). For industries where trust and provenance are crucial, such as food and pharmaceuticals, blockchain offers a way to ensure the integrity of products from origin to consumer. Advanced robotics and automation are also central to the concept of a smart SC. These technologies allow for the automation of repetitive tasks, such as picking, packing, and sorting, reducing the need for human intervention and minimizing the potential for errors. Robotics can operate continously without fatigue, leading to increased productivity and faster turnaround times. In warehouses and distribution centres, automation can significantly enhance operational efficiency and reduce labour costs (Mubarik and Khan 2024). The incorporation of these smart technologies into SC operations fundamentally transforms the way SCs function, creating what is known as a Smart SC Management (smart SCM). Smart SCM can be defined as a digitally enhanced, data-driven system that integrates advanced technologies to optimize and automate the flow of goods, information, and finances across the SC. It is characterized by its ability to process and respond to real-time data, predict and mitigate risks, and continuously improve efficiency and performance through self-learning algorithms.

Smart SCM is not just about adopting individual technologies; it involves designing an interconnected ecosystem where these technologies work together to create a more agile, efficient, and resilient SC. Various designs of smart SCs can be tailored to meet the specific needs and challenges of different industries. For instance, a SC in the manufacturing sector might focus on predictive maintenance and automated production lines, while a SC in the retail sector might prioritize demand forecasting, inventory optimization, and personalized customer experiences (Kandarkar and Ravi 2024). One example of a smart SC design is the use of IoT and AI in predictive maintenance within the automotive industry. Companies like General Motors have implemented systems where IoT sensors are embedded in machinery and vehicles to continuously monitor their condition. AI algorithms analyse this data to predict when maintenance is needed, allowing companies to schedule repairs before a breakdown occurs. This not only prevents costly downtime but also extends the lifespan of equipment, reducing overall maintenance costs. Another example is the use of blockchain and IoT in the food SC. Walmart, for instance, has adopted blockchain technology to enhance the traceability of its food products. By integrating IoT devices that track the location and condition of products at every stage of the SC,

Walmart can quickly trace the source of any food safety issues, such as contamination or spoilage. This design not only improves food safety but also boosts consumer confidence by providing transparent and reliable information about the origin and handling of their food. These examples illustrate the diverse ways in which smart SCs can be designed to meet the unique needs of different industries. As companies continue to explore and implement these designs, they are discovering new ways to overcome the challenges of traditional SCs and unlock new levels of efficiency and innovation.

1.4 Smart Supply Chain: Is It a Remedy to Modern Challenges?

Smart SCs offer a promising remedy for the myriad challenges facing traditional SCs, including issues of sustainability, resilience, and performance. By integrating advanced digital technologies such as IoT, AI, big data analytics, and blockchain, smart SCs can provide real-time visibility, predictive capabilities, and automated decision-making processes. These technologies enable companies to anticipate disruptions, optimize resource use, and adapt quickly to changes in the market, thereby reducing waste, enhancing efficiency, and improving overall SC performance. As a result, smart SCs are better equipped to handle the complexities and uncertainties of the modern business environment, positioning them as a key solution to the challenges that have plagued traditional SC models.

In the following paragraph, we explain as to how smart SC contributes to three specific SC challenges.

1.4.1 Smart SC and Sustainability Challenge

One of the most significant benefits of a smart SC is its ability to enhance the sustainability of a firm's operations. Sustainability in SC management involves minimizing the environmental impact of SC activities, ensuring ethical practices, and promoting the long-term viability of resources. Smart SCs achieve these goals through several mechanisms. First, the use of IoT and real-time data analytics allows companies to monitor and manage their resource consumption more effectively. For instance, sensors can track energy usage, water consumption, and emissions throughout the SC, providing valuable data that can be used to identify inefficiencies and implement corrective actions. This real-time monitoring enables companies to reduce their environmental footprint by optimizing resource use and minimizing waste (Movahed et al. 2024). AI and machine learning also play a crucial role in enhancing SC sustainability. These technologies can analyse large datasets to identify patterns and trends that might not be apparent through traditional methods. For example,

AI can optimize transportation routes to reduce fuel consumption and emissions, or it can predict demand more accurately to prevent overproduction and minimize waste. By making SCs more efficient and reducing unnecessary resource use, smart technologies contribute directly to sustainability efforts.

Blockchain technology further supports sustainability by improving transparency and traceability in the SC. With blockchain, every transaction and movement of goods is recorded in a secure, immutable ledger, making it easier to verify the origin and ethical sourcing of materials. This is particularly important in industries where sustainable sourcing is critical, such as food, fashion, and electronics (Mubarik and Shahbaz 2023). By ensuring that all components of a product are sourced responsibly, companies can meet consumer demands for ethical products and reduce the risk of reputational damage due to unsustainable practices. Moreover, smart SCs can support circular economy initiatives, which aim to reduce waste by reusing and recycling materials. For example, IoT sensors can track the condition of products throughout their lifecycle, enabling companies to reclaim and refurbish products at the end of their use, rather than discarding them. This not only reduces waste but also creates new business opportunities in the form of secondary markets for refurbished goods.

The above discussion gives a brief snapshot on as to how smart SCs can enhance SC sustainability.

1.4.2 Smart Supply Chain and Resilience Challenge

In addition to sustainability, smart SCs also significantly improve the resilience of SCs across three key facets: readiness, response, and recovery. SC resilience refers to the ability of a SC to anticipate, absorb, and recover from disruptions, whether they are caused by natural disasters, geopolitical tensions, or market fluctuations. Smart SCs enhance readiness by providing real-time visibility into the entire SC, from raw material suppliers to end consumers. IoT sensors, for instance, can continuously monitor SC conditions, such as inventory levels, production rates, and transportation status. This real-time data allows companies to detect potential issues before they escalate into major disruptions. For example, if a sensor detects a delay in the delivery of critical components, the system can automatically alert SC managers, who can then take proactive measures to mitigate the impact, such as rerouting shipments or finding alternative suppliers (Khan et al. 2023). AI and predictive analytics also play a crucial role in improving readiness by forecasting potential disruptions based on historical data and current trends. These technologies can identify patterns that indicate an increased risk of disruption, such as supplier financial instability, geopolitical tensions, or natural disaster likelihood. Armed with these insights, companies can develop contingency plans and take preventive actions, such as diversifying their supplier base or increasing inventory buffers in vulnerable regions.

When it comes to response, smart SCs enable faster and more effective reactions to disruptions. AI-driven decision-making systems can quickly analyse the situation, evaluate multiple response scenarios, and recommend the best course of action. For instance, if a natural disaster disrupts a key supplier, the system can immediately identify alternative suppliers, adjust production schedules, and reroute shipments to minimize downtime (Kusi-Saprong et al. 2024). This rapid SC response capability is critical in minimizing the impact of disruptions and maintaining continuity of operations. Blockchain technology also enhances the response by providing a transparent and immutable record of transactions, which can be invaluable in crisis situations where trust and verification are crucial. For example, in the event of a food safety recall, blockchain can quickly trace the affected products back through the SC, allowing companies to identify and isolate the source of the problem with minimal disruption to the rest of the SC. Recovery, the third facet of SC resilience, is also improved by smart SCs. After a disruption has occurred, the ability to quickly return to normal operations is essential. Smart SCs support recovery by leveraging AI and machine learning to optimize the post-disruption recovery process. For example, AI can help prioritize the allocation of limited resources, such as raw materials or production capacity, to the most critical areas of the SC, ensuring a faster return to full operational capacity. Additionally, the data collected during the disruption can be analysed to improve future resilience, helping the SC to learn and adapt from each event.

The above discussion highlights that smart SCs can enhance SC resilience by improving readiness through real-time monitoring and predictive analytics, enabling faster and more effective responses to disruptions, and supporting efficient recovery processes environment.

1.4.3 Smart Supply Chain and Smart Performance

Smart SC can have profound influence over SC performance. A smart SC, by definition, is designed to optimize performance across all key metrics, including cost, speed, quality, and customer satisfaction. By leveraging advanced technologies such as AI, IoT, and big data analytics, smart SCs can significantly enhance operational efficiency, reduce costs, and improve service levels (Kandarkar and Ravi 2024). One of the primary ways in which smart SCs improve performance is through enhanced visibility and data-driven decision-making. With real-time data flowing from every part of the SC, managers have a comprehensive view of operations and can make informed decisions quickly. This visibility allows for more accurate demand forecasting, inventory optimization, and production planning, all of which contribute to improved performance. For instance, AI-driven demand forecasting can help companies align production schedules more closely with actual demand, reducing the risk of overproduction and excess inventory, which in turn lowers costs and increases profitability. Another key area where smart SCs boost performance is in logistics and transportation. IoT sensors and advanced analytics enable companies to optimize transportation routes, reduce fuel consumption, and minimize delivery times (Khan et al. 2023). This not only reduces operational costs but also improves customer satisfaction by ensuring faster and more reliable deliveries. Additionally, automation and robotics in warehouses and distribution centres streamline order fulfilment processes, reducing lead times and improving accuracy. Furthermore, smart SCs enhance performance by enabling greater agility and flexibility. In today's rapidly changing market environment, the ability to quickly adapt to shifts in demand, supply, or market conditions is critical. Smart SCs, with their real-time monitoring and AI-driven analytics, can rapidly adjust to these changes, ensuring that the SC remains aligned with business objectives and market needs. For example, if a sudden spike in demand occurs, the smart SC can automatically adjust production levels, allocate additional resources, and expedite shipping to meet customer expectations.

In condensed form, smart SCs lead to smart SC performance by enhancing visibility, optimizing operations, reducing costs, and improving agility. The integration of advanced technologies into SC processes not only addresses the challenges of traditional SCs but also creates a competitive advantage for companies that adopt these innovations. As businesses continue to navigate the complexities of the global market, the ability to implement and leverage smart SCs will be a key determinant of success.

1.5 Concluding Remarks

This chapter sets out to provide a comprehensive introduction to the concept of smart SCM, exploring its definition, importance, and potential to address the pressing challenges faced by modern SCs. The focus was on understanding how smart technologies can transform traditional SCs into more resilient, sustainable, and high-performing systems, ultimately laying the foundation for a deeper exploration of smart SCM throughout the book.

We began by examining the current landscape of SCs, emphasizing how they have evolved into globally integrated networks that are critical to the success of modern businesses. These networks, while highly efficient, are also increasingly vulnerable to a range of unprecedented challenges, including natural disasters, geopolitical tensions, and demand for sustainability. The chapter highlighted the growing complexity of managing such vast and interconnected systems, where disruptions in one part of the world can ripple across the entire SC, causing significant delays, increased costs, and operational risks. To address these challenges, the chapter introduced the concept of smart SCM, defined as a digitally enhanced, data-driven system that leverages advanced technologies to optimize and automate SC operations. We explored the various smart technologies that are driving this transformation, including the Internet of Things (IoT), artificial intelligence (AI), machine learning, blockchain, and big data analytics. These technologies enable real-time monitoring, predictive analytics, and autonomous decision-making, creating SCs that are more agile, efficient, and capable of responding to disruptions. One of the key discussions in the chapter centred around how smart SCM can enhance the sustainability of SCs. By improving resource efficiency, reducing waste, and enabling transparency through blockchain, smart SCs help companies meet the growing demand for environmentally and socially responsible practices. We also discussed how these technologies support the circular economy by enabling the reuse and recycling of materials, thus reducing the environmental impact of SC activities. The chapter then turned to the role of smart SCM in enhancing SC resilience, focusing on the three critical facets of readiness, response, and recovery. We explained how real-time data, predictive analytics, and AI-driven decision-making allow companies to anticipate potential disruptions, respond quickly when they occur, and recover more efficiently afterwards. These capabilities are crucial for maintaining continuity in an increasingly volatile global environment, where traditional SC models often fall short. Finally, we examined the impact of smart SCM on SC performance, arguing that the integration of advanced technologies leads to smarter, more efficient operations. Enhanced visibility, optimized logistics, and greater agility were highlighted as key benefits that contribute to improved cost management, faster delivery times, and higher customer satisfaction. The ability to align SC operations more closely with business objectives and market demands positions companies that adopt smart SCM at a significant competitive advantage.

Looking forward, the way forward for smart SCM involves not only the continued adoption of these advanced technologies but also a strategic approach to their implementation. As organizations increasingly recognize the value of smart SCs, there will be a growing emphasis on integrating these technologies seamlessly into existing operations. This requires a thorough understanding of the specific needs and challenges of each SC, as well as careful planning and investment in the necessary infrastructure and skills. Moreover, as smart technologies continue to evolve, so too will the potential of smart SCM. Emerging trends such as quantum computing, next-generation AI, and advanced robotics promise to push the boundaries of what is possible in SCM. These innovations will further enhance the capabilities of smart SCs, enabling even greater levels of efficiency, resilience, and sustainability.

Acknowledgements We acknowledge the support of Higher Education Commission of Pakistan under NRPU: 20-11226 (PI: Dr. Shujaat Mubarik).

References

Agrawal, S., Agrawal, R., Kumar, A., Luthra, S., & Garza-Reyes, J. A. (2024). Can industry 5.0 technologies overcome supply chain disruptions?—a perspective study on pandemics, war, and climate change issues. *Operations Management Research, 17*(2), 453-468.

AlMulhim, A. F. (2021). Smart supply chain and firm performance: the role of digital technologies. *Business Process Management Journal, 27*(5), 1353-1372.

Boudouaia, M. A., Ouchani, S., Qaisar, S. M., & Almaktoom, A. T. (2024, January). Supply Chain 5.0: Vision, Challenges, and Perspectives. In *2024 21st Learning and Technology Conference (L&T)* (pp. 203–208). IEEE.

Fan, D., Zhou, Y., Yeung, A. C., Lo, C. K., & Tang, C. (2022). Impact of the US–China trade war on the operating performance of US firms: The role of outsourcing and supply base complexity. *Journal of Operations Management, 68*(8), 928-962.

Ivanov, D., & Dolgui, A. (2021). A digital supply chain twin for managing the disruption risks and resilience in the era of Industry 4.0. *Production Planning & Control, 32*(9), 775–788.

Ivanov, D., Dolgui, A., & Sokolov, B. (2019). The impact of digital technology and Industry 4.0 on the ripple effect and supply chain risk analytics. *International journal of production research, 57*(3), 829-846.

Kandarkar, P. C., & Ravi, V. (2024). Investigating the impact of smart manufacturing and interconnected emerging technologies in building smarter supply chains. *Journal of Manufacturing Technology Management.*

Khan, S. A., Gupta, H., Gunasekaran, A., Mubarik, M. S., & Lawal, J. (2023). A hybrid multi-criteria decision-making approach to evaluate interrelationships and impacts of supply chain performance factors on pharmaceutical industry. *Journal of Multi-Criteria Decision Analysis, 30*(1–2), 62–90.

Khan, S. A., Mubarik, M. S., Zaman, S. I., Ishizaka, A., & Saleem, K. (2023). Transparency enhancement of supply chain capabilities through social media: a proposed hierarchical model. *Environment, Development and Sustainability,* 1–39.

Kusi-Sarpong, S., Mubarik, M. S., & Khan, S. A. (2024). Supply chain mapping for "visilience": role of blockchaindriven supply chain management. In *The Palgrave Handbook of Supply Chain Management* (pp. 585–599). Cham: Springer International Publishing.

Lee, K. L., Wong, S. Y., Alzoubi, H. M., Al Kurdi, B., Alshurideh, M. T., & El Khatib, M. (2023). Adopting smart supply chain and smart technologies to improve operational performance in manufacturing industry. *International Journal of Engineering Business Management, 15,* 18479790231200614.

Movahed, A. B., Movahed, A. B., & Nozari, H. (2024). Opportunities and challenges of smart supply chain in Industry 5.0. *Information Logistics for Organizational Empowerment and Effective Supply Chain Management,* 108–138.

Mubarik, M. S., & Shahbaz, M. (2023). *Blockchain Driven Supply Chain Management.* Springer: Singapore (ISSN 2192-810X).

Mubarik, M. S., & Khan, S. A. (2024). *The Theory, Methods and Application of Managing Digital Supply Chains.* Emerald Publishing Limited.

Mubarik, M. S., Naghavi, N., Mubarik, M., Kusi-Sarpong, S., Khan, S. A., Zaman, S. I., & Kazmi, S. H. A. (2021). Resilience and cleaner production in industry 4.0: role of supply chain mapping and visibility. *Journal of cleaner production, 292,* 126058.

Shi, X., Liu, W., & Zhang, J. (2023). Present and future trends of supply chain management in the presence of COVID-19: a structured literature review. *International Journal of Logistics Research and Applications, 26*(7), 813-842.

Vahidinia, A., & Hasani, A. (2023). A comprehensive evaluation model for smart supply chain based on the hybrid multi-criteria decision-making method. *Journal of Soft Computing and Decision Analytics, 1*(1), 219-237.

Zhang, H., Aydin, G., & Parker, R. P. (2022). Social responsibility auditing in supply chain networks. *Management Science, 68*(2), 1058–1077.

Zhang, G., Yang, Y., & Yang, G. (2023). Smart supply chain management in Industry 4.0: the review, research agenda and strategies in North America. *Annals of operations research, 322*(2), 1075-1117.

Zhu, A., Han, Y., & Liu, H. (2024). Effects of adaptive cooperation among heterogeneous manufacturers on supply chain viability under fluctuating demand in post-COVID-19 era: an agent-based simulation. *International Journal of Production Research, 62*(4), 1162-1188.

Chapter 2
Artificial Intelligence for Smart Supply Chain Management: Opportunities and Challenges

Md. Ramjan Ali and Shah Md. Ashiquzzaman Nipu

2.1 Introduction

Supply chains have continuously faced various challenges, necessitating periodic upgrades to maintain efficiency and competitiveness. Each era has presented unique obstacles that have required innovative solutions to ensure the seamless movement of goods, information, and funds (Khan et al. 2022a, b, c; Mubarik et al. 2023). Historically, supply chains have evolved to address these challenges, fostering collaboration and communication among partners. In the current era, however, the challenges are unprecedented and multifaceted, driven by factors such as globalization, rapid technological advancements, and heightened consumer expectations. These unique pressures demand a radical transformation of traditional supply chains. Digital technologies (DTs) have emerged as pivotal opportunities in this context. Technologies like AI, IoT, and blockchain offer new avenues for intelligent decision-making, real-time visibility, and enhanced efficiency. Historically, technological advancements have played crucial roles in improving supply chain resilience and responsiveness. For instance, in the 1990s and 2000s, the introduction of barcode scanners for inventory tracking and Electronic Data Interchange (EDI) for information sharing marked significant milestones in automating supply chain processes (Khayyam et al. 2022). These innovations laid the groundwork for the smart supply chain (SSC), which integrates emerging technologies at every stage.

In the present era, AI stands out as a transformative technology with the potential to revolutionize supply chains. AI encompasses machine learning, natural language processing, deep learning, and computer vision, optimizing each process and step

Md. R. Ali (✉) · S. Md. A. Nipu
Department of Mechanical and Production Engineering, Ahsanullah University of Science and Technology, Dhaka, Bangladesh
e-mail: aliramjan808@gmail.com

S. Md. A. Nipu
e-mail: nipu.ipe@aust.edu

M. S. Mubarik and S. A. Khan (eds.), *Smart Supply Chain Management*,
https://doi.org/10.1007/978-981-96-1333-5_2

within the supply chain (Holzinger et al. 2023). By offering human-level intelligence, AI is reshaping various industries, including supply chain management (Na et al. 2022). AI aids in traffic management, autonomous vehicles, product recommendations, customer service, fraud detection, automated production, quality control, and inventory management. Its ability to manage vast amounts of data, identify patterns, and make predictions enables businesses to automate their supply chain processes, from procurement to logistics. Despite its potential, AI adoption in supply chains faces several challenges (Ali et al. 2023). Integrating AI with traditional supply chain processes is complex, requiring significant investments and skilled personnel (Kehayov et al. 2022; Naz et al. 2022). Additionally, data security and ethical considerations pose significant hurdles. For instance, AI-driven automation can lead to job displacement, impacting economies (Su et al. 2022). Moreover, data sharing reluctance among supply chain members can hinder the creation of a fully integrated SSC. Nevertheless, some companies have successfully harnessed AI to enhance their supply chains. Case studies highlight instances where AI improved sales configuration, production planning, quality inspection, and risk management (Helo and Hao 2022; Shah et al. 2021; Ganesh and Kalpana 2022). These examples demonstrate AI's capability to provide competitive advantages, increase profit margins, and reduce risks. However, AI adoption remains relatively low, with many applications still in their infancy (Baryannis et al. 2019).

This chapter aims to fill the gaps in the literature by exploring the role of AI in SSCs. It addresses how AI can improve supply chain resilience, sustainability, and responsiveness. Furthermore, the chapter highlights the enablers and barriers to AI adoption in SSCs, providing insights for industry experts to enhance adoption rates and dispel misconceptions. The chapter also outlines future developments and trends for SSCs.

The structure of the chapter is as follows: Sect. 2.2 discusses the motivation for this study. Section 2.3 provides an overview of SSCs. Section 2.4 offers an overview of AI. Section 2.5 explores the relationship between AI and SSCs. Section 6 examines the enablers and barriers to AI implementation in SSCs. Section 2.7 presents discussions and implications. Section 2.8 concludes the chapter.

2.2 Background

The adoption of AI by companies has surged, leading to more efficient business operations within their supply chains (Alzarooni et al. 2022). This trend is driven by the diverse demands of customers, prompting companies to leverage digital technologies. AI, in particular, has shown significant potential in enhancing customer satisfaction. For instance, Amazon, the world's largest e-commerce platform, uses machine learning to optimize warehouse and logistics operations (Dash et al. 2019). By utilizing AI technologies such as robotics, predictive analysis, and automated sorting systems, Amazon has reduced costs and delivery times while also improving customer service through digital chatbots.

Similarly, Walmart has embedded AI into its supply chain to forecast demand, anticipate consumer needs, and maintain inventory efficiently (Praveen et al. 2019). Maersk, the largest container shipping company in the world, has implemented AI to optimize shipping routes and establish a real-time cargo tracking and monitoring system (Attaran 2020; Wang and Sarkis 2021). Toyota has fully integrated AI into their manufacturing processes to ensure quality and predictive maintenance, thereby lowering supply chain costs (Javaid et al. 2022). Nestle, a leading FMCG company, uses AI for its advertising campaigns (Sabry Riad Abdel Wanes 2021).

While large companies are reaping the rewards of AI, smaller companies are still debating its applicability and potential value in their industries. This acts as an impetus for the present research leading to discuss as to what role AI can play in smart SC irrespective of size of a company.

2.3 Smart Supply Chain: An Overview

The core focus of supply chain management (SCM) is ensuring the right product is available at the right time, place, quantity, and price (Barreto et al. 2017; Susan 2010). Traditional supply chains often face challenges like overstocking, stockouts, and delays due to their complexity and uncertainty. To address these issues, supply chains are evolving into smart supply chains (SSCs) by integrating logistics, business, information, and financial flows. SSCs utilize modern technologies such as AI, IoT, blockchain, and cloud computing to improve cost efficiency and productivity. Modern SSCs are being rapidly developed and implemented because they offer significant benefits, such as improved decision-making through data analysis, refined business processes, and reduced inefficiencies (Cioffi et al. 2020; Zhang et al. 2023). As supply chains become more complex and interconnected, the risk of disruptions increases (Abdel-Basset 2019; Khan et al. 2021). Traditional supply chain strategies often fall short in addressing these challenges. Therefore, integrating intelligence into decision-making and management tools is crucial (Granell et al. 2020). AI and big data analytics enhance processes, reduce waste, and improve decision-making in SSCs. Transparency is also a key feature, as it allows all parties to access vital data, fostering trust and speeding up problem-solving. SSCs are resilient, enabling businesses to prevent issues through real-time tracking and predictive data analysis. Moreover, SSCs can incorporate ecological considerations, using blockchain and IoT to ensure sustainable sourcing, meeting the demand for environmentally friendly products, and boosting brand value.

2.4 Artificial Intelligence (AI) and Supply Chain

The supply chain logistics field is undergoing a significant technological transition with the integration of AI and IoT. AI enhances supply chain management by facilitating intelligent activities such as knowledge acquisition, problem-solving, decision-making, and natural language processing. For example, Fig. 2.1 illustrates how AI integrates supply chain major components such as suppliers, manufacturers, logistics providers, warehouses, and customers. With the assistance of AI, it is possible to predict customer demands and other entities can take action based on this prediction. Additionally, this integration improves transparency, agility, and adaptability, addressing various supply chain challenges (Nozari et al. 2022a, b). AI and IoT are seamlessly incorporated into modern operations, contributing to industrial automation through the integration of factory command networks, wireless identification systems, information management systems, and sensor networks. These technologies impact supply chain functions such as sourcing, logistical planning, inventory management, and vendor selection, optimizing efficiency and reducing costs. AI and machine learning enhance basic business strategies and operations, while IoT improves communication and data sharing in real-time, benefiting both internal and external organizational interactions. This synergy has attracted the attention of researchers and industry professionals. Research shows AI's broad applicability across various sectors, including medicine, business, education, and entertainment. Ethical considerations, such as privacy, transparency, and bias, are crucial for responsible AI adoption. Collaboration between academics, policymakers, and the public is necessary to address these implications. Studies have also explored AI's impact on the agri-food supply chain (Treiblmaier et al. 2020).

2.5 Artificial Intelligence and Smart Supply Chain

AI and smart supply chains (SSCs) are revolutionizing how organizations manage daily operations. AI replicates human intelligence in machines, while SSCs integrate technology to enhance supply chain efficiency, visibility, and adaptability. This integration, facilitated by technologies like machine learning, big data analytics, IoT, and blockchain, is driving significant improvements in supply chain management (Benzidia et al. 2021). AI improves supply chain operations by optimizing inventory management, demand forecasting, and route planning (Zhang et al. 2021; Di Vaio et al. 2020). AI can analyze historical data to predict future demand, enhancing inventory management and reducing costs. It also provides real-time visibility, allowing organizations to monitor goods from origin to consumer, quickly identifying and resolving issues, which reduces delays and increases customer satisfaction (Dubey et al. 2021).

Moreover, AI enhances supply chain resilience by using predictive analytics to identify and mitigate potential risks, such as natural disasters (Baryannis et al.

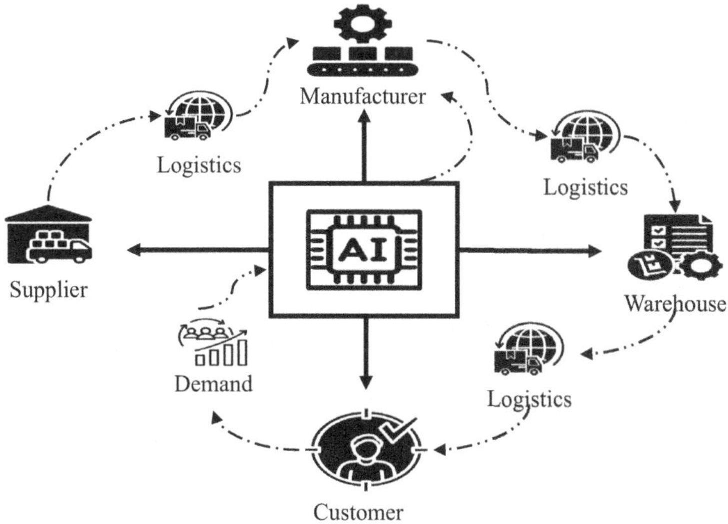

Fig. 2.1 Contribution of AI in supply chain

2019). It also fosters collaboration within the supply chain by improving communication among consumers, distributors, and suppliers, thereby reducing lead times and increasing efficiency (Min 2010). In logistics, AI opens new avenues for efficiency and productivity through predictive analytics and automation (Gupta et al. 2022). AI algorithms analyze data from various sources to forecast supply chain bottlenecks (Shankar et al. 2023; Younis et al. 2022), allowing logistics managers to make data-driven decisions and avoid costly delays (Amirkolaii et al. 2017). AI-powered robots and drones can automate tasks such as order picking, packing, and transportation, enhancing speed, accuracy, and reducing labor costs (Bergener et al. 2019).

AI also transforms planning processes across industries by analyzing vast amounts of data to optimize decisions regarding inventory, personnel, and production schedules (Toorajipour et al. 2021). AI can identify patterns and anticipate future events, improving overall planning efficiency. Additionally, it helps in risk management by evaluating data to foresee supply chain disruptions or financial risks (Baryannis et al. 2019). The symbiotic relationship between AI and SSCs further enhances the capabilities of both. SSCs, through technologies like sensors and IoT, capture real-time data that trains AI systems for better predictions and recommendations (Dauvergne 2022; Khan et al. 2023). This continuous learning loop improves AI's decision-making precision, benefiting supply chain operations over time (Los 2022).

2.6 Implementation of Artificial Intelligence in Smart Supply Chain: Enablers and Barriers

From the aforementioned discussion, it is clear that AI can improve SSC, and still many industries facing many challenges to adopting it. Hence, it is necessary to find out the factors which can enable and hinder the implementation of AI in SSC to enhance the adoption rate. Most of the enablers and barriers which are identified in previous studies are extracted to represent them in a structured manner. To accomplish this, the keywords "enablers to implementing artificial intelligence in smart supply chain, barriers to implementing artificial intelligence in smart supply chain, enablers of AI and smart supply chain, barriers to implementing artificial intelligence in smart supply chain, AI and smart supply chain enablers, AI and smart supply chain barriers" were used to extensively search in databases such as Scopus, Science Direct, Google Scholar, and Web of Science. There were no white papers, non-academic papers, or conference papers considered for this book chapter. Also, only academic papers written in English are accepted. Furthermore, the reference list of related studies is skimmed to ensure that most of the enablers and barriers are considered. At last, we found in total 22 articles which are concentrated on to describe barriers and drivers to implement AI. However, the enablers and barriers are mentioned in the following sub-section.

2.6.1 Enablers

In recent years, the use of AI in managing supply chains has become more significant. This is due to the fact that businesses are constantly striving to improve their operations and maintain their competitive edge in the dynamic and ever-shifting global market. The following are some of the most key drivers that firms may utilize in order to deploy AI in their smart supply chains successfully:

- Data quality and availability: The availability of data of high quality is one of the most essential enablers for putting AI to use in the management of supply chains (Attia et al. 2023; Edwin Cheng et al. 2022; Susan 2010). For successful decision-making that is enabled by AI, it is essential to have data that is accurate and up to date from a variety of sources, such as suppliers, customers, and internal systems. In order to accomplish this, organizations need to set data governance standards that assure the quality of the data, that it is full, and that it is consistent among the various systems and applications.
- Improvements in both operational effectiveness and output: SSCs, which are powered by AI, are significantly more productive and efficient than traditional supply chains. AI technology such as machine learning and predictive analytics may assist businesses in optimizing their inventory levels, minimizing waste, and cutting down on lead times (Attia et al. 2023; Sætra 2021). For instance, AI

systems, may examine data gleaned from sensors and other Internet of Thing's devices to devise optimal routes, shortening the time spent traveling.

- Talent and skills: Having the appropriate people with the relevant talents already working for you is another crucial enabler for integrating AI in supply chain management. It is necessary for businesses to employ data scientists, AI developers, and business analysts that are able to collaborate in order to design and implement AI-enabled solutions that can enhance the functioning of supply chain operations. Companies may create the capabilities necessary to exploit AI in their supply chains by enhancing the skills of their existing workforce as well as recruiting new talent (Butner 2010; Chen et al. 2022; Khan et al. 2021).
- Increased transparency and a clearer audit trail: Increased visibility and traceability across the supply chain may be realized with the help of AI in SSC. This enables firms to follow the movement of items from manufacture to delivery and detect possible bottlenecks in the supply chain. This increased visibility may also assist firms to improve their inventory management, which can reduce the risk of stockouts as well as the danger of overstocking (Rodríguez et al. 2018).
- Cost savings: There are several ways in which firms might save costs with the assistance of intelligent supply networks driven by AI. For instance, predictive analytics may be utilized to optimize inventory levels, minimizing the requirement for carrying excessive inventory. Algorithms employed in machine learning may also be used to optimize production schedules, which can cut down on the time and resources needed to produce things (Vázquez et al. 2020).
- Increased levels of contentment among customers: By offering more insight into the supply chain, smart supply chains that are powered by AI may assist businesses in increasing the level of customer satisfaction they experience. It is possible for this to assist enterprises in providing accurate delivery estimates and lowering the danger of stockouts. In addition, AI technology such as chatbots and virtual assistants may be utilized to offer real-time assistance and updates to clients (De Azambuja 2021; Di Vaio et al. 2020; Vázquez et al. 2020).
- Competitive advantage: In order to achieve a competitive edge, businesses may improve their operational efficiency and cut costs by using intelligent supply chains that are powered by AI (Rodríguez et al. 2018). In addition, businesses that are able to provide improved visibility and traceability throughout the entirety of the supply chain are in a position to differentiate themselves from their rivals.

The implementation of AI in smart supply chains is being driven by a variety of factors, including increased efficiency and productivity, improved decision-making, enhanced visibility and traceability, cost savings, increased customer satisfaction, competitive advantage, and regulatory compliance. It is probable that we will see increasing adoption of AI technology in the years to come as businesses continue to explore methods to enhance their supply chains. This is because companies are constantly looking for new ways to improve their supply chains. However, in the next sub-section, factors which are acting as barriers to implement AI in SSC are mentioned.

2.6.2 Barriers

AI is a revolutionary innovation with the potential to significantly advance the supply chain industry. The application of AI to supply chain management has the potential to have a number of positive outcomes, including increased operational effectiveness, decreased operational costs, increased visibility, etc. (De Azambuja 2021). Nevertheless, in spite of the numerous potential benefits, there are a number of obstacles that prevent the application of AI in smart supply chains. In the following sections, we will go through some of the barriers that must first be overcome in order to successfully use AI in a smart supply chain.

- *Quality of data and accessibility*: To train AI systems, a lot of high-quality data is needed. AI's predictions and choices depend on data quality (Khan et al. 2021; Reshad et al. 2023). Unfortunately, outdated systems in many supply chains generate data in disparate forms, making data cleaning, normalization, and integration problematic.
- *High investment:* AI application in SSCs is hindered by high development and implementation expenses (Kashem et al. 2023; Grimm et al. 2014; Hofmann et al. 2018; Meinlschmidt et al. 2018). AI algorithms and supply chain implementation need large hardware, software, and labor expenditures. Integrating AI into current systems is very complicated and requires specific skills.
- *Privacy and cybersecurity:* AI needs lots of data. This makes the technology vulnerable to hacks and privacy intrusions. Companies must safeguard their data from hackers (De Azambuja 2021; Grimm et al. 2014; Hofmann et al. 2018; Reshad et al. 2023).
- *Lack of trust:* AI technology's lack of trust hinders its use in smart supply networks. Many companies are wary of using AI algorithms to make judgments out of concern that they may be prejudiced or inaccurate (Dou et al. 2018; Hofmann et al. 2018). Employees may oppose AI technologies due to a lack of confidence.
- *Human capital shortage*: AI design, development, and implementation demand specific talents (Reshad et al. 2023). AI professionals are scarce. Thus, AI system development and management skills may be scarce.
- *Integrating legacy systems*: Integrating AI with legacy systems is another SSC AI difficulty. Many firms have invested much in their present systems, making new technology integration difficult and time-consuming. AI may need system replacement in some circumstances (Kembro et al. 2017; Sarkis et al. 2019).
- *Change resistant:* Supply chain AI demands a major culture change. Change aversion can hinder AI adoption. New technology may be unpopular or beyond employees' abilities (Nejati et al. 2017).
- *Legal and regulatory obstacles*: Finally, smart supply chain AI faces legal and regulatory hurdles. As AI improves, authorities are taking notice. AI systems must comply with rules and regulations, which may be difficult and time-consuming (Cubric 2020; De Azambuja 2021).

Overall, implementing AI in the supply chain is complex. Addressing such obstacles can improve business efficiency, cost, and decision-making policy. To overcome these challenges, high-quality data, AI integration with legacy systems, experienced workers, and cybersecurity and privacy concerns must be addressed. A well-planned, all-encompassing strategy is required to successfully implement AI in the supply chain.

2.7 Implications

This chapter highlights some implications for business managers. From the above discussion, it can be said that AI and SSC can bring a revolutionary change in the way of business operations. This will enable business managers to make decisions based on their experience as well as previous data, allowing them to make more appropriate decisions in uncertain situations. However, implementing AI in SSC can require a considerable amount of investment. Business managers must consider the variety of implications to weigh the advantages against the costs. In addition, AI-based SSC can require skilled personnel to operate efficiently. Due to this, business managers must invest in employee skill development by offering training. Further, adopting AI for SSC can require investments in technology which needs to be addressed by the business managers. Moreover, implementing AI in SSC can change the organizational structure and management practice. So, managers must carefully handle this for implementing and maintaining AI-SSC. However, sometimes AI algorithms can be modified to get individual advantages. Therefore, business managers must make the data and algorithms transparent to their collaborative parties for ethical considerations. Also, business managers must provide data security and ensure that the sensitive data is completely protective.

AI-based SSC offers entrepreneurs both opportunities and challenges. In some cases, the laws governing the use of AI vary between regions. This regulation must be observed by entrepreneurs and they must ensure that their compliance with other regulations is up to date (Baah et al. 2022). Entrepreneurs must therefore take into account the regulatory system for AI and SSC (Modgil et al. 2022). Also, entrepreneurs need to develop a strategic and thoughtful plan for implementing AI in SSC in order to maximize the benefits while minimizing the risks and challenges. In addition to considering the potential impact on employees and the broader community, entrepreneurs must also consider the possible benefits of AI-based SSC. Further, the integration of AI and SSC may result in workforce disruptions which can have negative impacts on employee morale, the local community, and country's economy. Entrepreneurs must have a plan to balance the workforce. It is also essential to make sure that AI will be used ethically and in a manner that protects individual rights and privacy. For this, policymakers must ensure that AI systems are transparent, as well as that decision-making processes are explicable, to promote accountability and prevent discrimination. Moreover, policymakers can aid shape the future of AI and SSC by creating a supportive regulatory environment and addressing the ethical, privacy, and workforce implications of AI. Policymakers can also promote that these technologies can benefit society and drive economic growth.

Again, AI implementation in SSC has also implications for academia and educational institutions since this provides an exciting area for researchers to explore more benefits of implementing AI, challenges, applications, etc. Also, educational institutions can help students to develop skills that are currently or in the future will require for the job market. Further, educational institutions should ensure that their curricula are inclusive and accessible. This will support students from diverse backgrounds to gain knowledge required to succeed in the increasingly automated and data-driven world. Students in the high school or college should be taught about the ethical issues that must be addressed in order to implement AI. Teachers should also encourage students to involve in research to identify potential risks, impacts, etc. associated with SSC. Additionally, academia and educational institutions have the opportunity to collaborate with industry and the government to advance the development and utilization of SSC with AI. Moreover, AI in the supply chain prompts the development of innovative tools and platforms for continuous improvement. It is possible for academic researchers to contribute to the development of these tools and to develop appropriate solutions based on their knowledge. Additionally, it will be easier for entrepreneurs to achieve workforce balance if educational institutions can equip students with the skills they need to deal with AI.

2.8 Conclusion

This chapter explored the transformative role of artificial intelligence (AI) in smart supply chains (SSCs), highlighting the opportunities and challenges that accompany this integration. The primary objectives were to assess how AI can reduce supply chain costs, enhance transparency, and protect against disruptions and fraud, as well as evaluate the role of AI in automating manufacturing processes and managing inventory efficiently. Key findings indicate that AI is pivotal in offering significant advantages for SSCs, such as enhanced decision-making, cost reduction, improved environmental sustainability, and overall supply chain resilience. AI facilitates task automation and optimizes inventory management, making supply chains more efficient, resilient, and adaptive to changing conditions. AI and SSCs are revolutionizing how organizations manage their daily operations. AI replicates human intelligence in machines, enabling intelligent decision-making that enhances the effectiveness, visibility, and adaptability of supply chain management. SSCs integrate technologies like machine learning, big data analytics, IoT, and blockchain to improve cost efficiency and productivity. AI optimizes inventory management, demand forecasting, and route planning by analyzing historical data, predicting future demand, and improving inventory management to reduce costs. AI also provides real-time visibility into supply chains, allowing organizations to monitor goods from origin to consumer, swiftly detect and resolve issues, and reduce delays, increasing customer satisfaction.

AI-driven SSCs offer numerous benefits, including improved efficiency and productivity, enhanced visibility and traceability, cost savings, increased customer satisfaction, resilience and risk management, improved collaboration, and ecological sustainability (Kashem et al. 2023; Kazancoglu et al. 2023). Despite these advantages, several challenges hinder AI adoption in supply chains, such as data quality and availability, high investment costs, privacy and cybersecurity concerns, lack of trust and ethical issues, skilled workforce shortages, integration with legacy systems, and regulatory and legal challenges.

Businesses must weigh the advantages of AI against the costs and ensure that skilled personnel are available to operate AI-based systems. Investment in technology and employee training is crucial for successful AI integration. Managers must also handle changes in organizational structure and management practices carefully to maintain efficiency and morale. Policymakers play a crucial role in shaping the regulatory environment for AI, ensuring that AI systems are transparent and accountable while protecting individual rights and privacy. Future research should focus on identifying and addressing barriers to AI adoption in different industries and scenarios, exploring hidden applications of AI in SSCs to reveal new opportunities for improvement. Researchers should continue investigating the ethical, social, and economic implications of AI to ensure responsible adoption.

References

Abdel-Basset, M. (2019). Corrigendum to "A framework for risk assessment, management and evaluation: Economic tool for quantifying risks in supply chain" (Future Generation Computer Systems (2019) 90 (489–502), (S0167739X18312172) (https://doi.org/10.1016/j.future.2018.08.035)). *Future Generation Computer Systems, 93*, 1076–1077. https://doi.org/10.1016/j.future.2018.12.056

Ali, M. R., Nipu, S. M. A., & Khan, S. A. (2023). A decision support system for classifying supplier selection criteria using machine learning and random forest approach. *Decision Analytics Journal*, 100238.

Alzarooni, A. M., Khan, S. A., Gunasekaran, A., & Mubarik, M. S. (2022). Enablers for digital supply chain transformation in the service industry. *Annals of Operations Research*. https://doi.org/10.1007/s10479-022-05047-x

Amirkolaii, K. N., Baboli, A., Shahzad, M. K., & Tonadre, R. (2017). Demand Forecasting for Irregular Demands in Business Aircraft Spare Parts Supply Chains by using Artificial Intelligence (AI). *IFAC-PapersOnLine, 50*(1), 15221–15226. https://doi.org/10.1016/j.ifacol.2017.08.2371

Attaran, M. (2020). Digital technology enablers and their implications for supply chain management. *Supply Chain Forum: An International Journal, 21*(3), 158–172. Taylor & Francis.

Attia, E. A., Alarjani, A., Uddin, M. S., & Kineber, A. F. (2023). Determining the Stationary Enablers of Resilient and Sustainable Supply Chains. *Sustainability (Switzerland), 15*(4). https://doi.org/10.3390/su15043461

Baah, C., Acquah, I. S. K., & Ofori, D. (2022). Exploring the influence of supply chain collaboration on supply chain visibility, stakeholder trust, environmental and financial performances: a partial least square approach. *Benchmarking: An International Journal, 29*(1), 172–193.

Barreto, L., Amaral, A., & Pereira, T. (2017). Industry 4.0 implications in logistics: an overview. *Procedia Manufacturing, 13*, 1245–1252.

Baryannis, G., Validi, S., Dani, S., & Antoniou, G. (2019). Supply chain risk management and artificial intelligence: state of the art and future research directions. *International Journal of Production Research, 57*(7), 2179–2202.

Benzidia, S., Makaoui, N., & Bentahar, O. (2021). The impact of big data analytics and artificial intelligence on green supply chain process integration and hospital environmental performance. *Technological Forecasting and Social Change, 165*(December 2020), 120557. https://doi.org/10.1016/j.techfore.2020.120557

Bergener, K., Räckers, M., & Stein, A. (2019). The art of structuring: bridging the gap between information systems research and practice. In *The Art of Structuring: Bridging the Gap between Information Systems Research and Practice*. Springer International Publishing. https://doi.org/10.1007/978-3-030-06234-7

Butner, K. (2010). The smarter supply chain of the future. *Strategy and Leadership, 38*(1), 22–31. https://doi.org/10.1108/10878571011009859

Chen, P.-K., He, Q.-R., & Chu, S. (2022). Influence of blockchain and smart contracts on partners' trust, visibility, competitiveness, and environmental performance in manufacturing supply chains. *Journal of Business Economics and Management, 23*(4), 754–772.

Cioffi, R., Travaglioni, M., Piscitelli, G., Petrillo, A., & De Felice, F. (2020). Artificial intelligence and machine learning applications in smart production: Progress, trends, and directions. *Sustainability (Switzerland), 12*(2). https://doi.org/10.3390/su12020492

Cubric, M. (2020). Drivers, barriers and social considerations for AI adoption in business and management: A tertiary study. *Technology in Society, 62*, 101257.

Dash, R., McMurtrey, M., Rebman, C., & Kar, U. K. (2019). Application of artificial intelligence in automation of supply chain management. *Journal of Strategic Innovation and Sustainability, 14*(3), 43–53.

De Azambuja, L. S. (2021). Drivers and Barriers for the development of Smart Sustainable Cities: A Systematic Literature Review. *ACM International Conference Proceeding Series*, 422–428. https://doi.org/10.1145/3494193.3494250

Dauvergne, P. (2022). Is artificial intelligence greening global supply chains? Exposing the political economy of environmental costs. *Review of International Political Economy, 29*(3), 696–718. https://doi.org/10.1080/09692290.2020.1814381

Di Vaio, A., Palladino, R., Hassan, R., & Escobar, O. (2020). Artificial intelligence and business models in the sustainable development goals perspective: A systematic literature review. *Journal of Business Research, 121*(August), 283–314. https://doi.org/10.1016/j.jbusres.2020.08.019

Dou, Y., Zhu, Q., & Sarkis, J. (2018). Green multi-tier supply chain management: An enabler investigation. *Journal of Purchasing and Supply Management, 24*(2), 95–107. https://doi.org/10.1016/j.pursup.2017.07.001

Dubey, R., Bryde, D. J., Blome, C., Roubaud, D., & Giannakis, M. (2021). Facilitating artificial intelligence powered supply chain analytics through alliance management during the pandemic crises in the B2B context. *Industrial Marketing Management, 96*(February), 135–146. https://doi.org/10.1016/j.indmarman.2021.05.003

Edwin Cheng, T. C., Kamble, S. S., Belhadi, A., Ndubisi, N. O., Lai, K. hung, & Kharat, M. G. (2022). Linkages between big data analytics, circular economy, sustainable supply chain flexibility, and sustainable performance in manufacturing firms. *International Journal of Production Research, 60*(22), 6908–6922. https://doi.org/10.1080/00207543.2021.1906971

Ganesh, A. D., & Kalpana, P. (2022). Future of artificial intelligence and its influence on supply chain risk management–A systematic review. *Computers & Industrial Engineering*, 108206.

Granell, C., Kamilaris, A., Kotsev, A., Ostermann, F. O., & Trilles, S. (2020). Internet of things. *Manual of Digital Earth*, 387–423.

Grimm, J. H., Hofstetter, J. S., & Sarkis, J. (2014). Critical factors for sub-supplier management: A sustainable food supply chains perspective. *International Journal of Production Economics, 152*, 159–173. https://doi.org/10.1016/j.ijpe.2013.12.011

Gupta, H., Yadav, A. K., Kusi-Sarpong, S., Khan, S. A., & Sharma, S. C. (2022). Strategies to overcome barriers to innovative digitalisation technologies for supply chain logistics resilience during pandemic. *Technology in Society*, *69*, 101970.

Helo, P., & Hao, Y. (2022). Artificial intelligence in operations management and supply chain management: an exploratory case study. *Production Planning and Control*, *33*(16), 1573–1590. https://doi.org/10.1080/09537287.2021.1882690

Hofmann, H., Schleper, M. C., & Blome, C. (2018). Conflict Minerals and Supply Chain Due Diligence: An Exploratory Study of Multi-tier Supply Chains. *Journal of Business Ethics*, *147*(1), 115–141. https://doi.org/10.1007/s10551-015-2963-z

Holzinger, A., Keiblinger, K., Holub, P., Zatloukal, K., & Müller, H. (2023). AI for life: Trends in artificial intelligence for biotechnology. *New Biotechnology*.

Javaid, M., Haleem, A., Singh, R. P., Rab, S., Suman, R., & Khan, S. (2022). Exploring relationships between Lean 4.0 and manufacturing industry. *Industrial Robot: The International Journal of Robotics Research and Application*, *49*(3), 402–414.

Kashem, M. A., Shamsuddoha, M., Nasir, T., & Chowdhury, A. A. (2023). Supply Chain Disruption versus Optimization: A Review on Artificial Intelligence and Blockchain. *Knowledge*, *3*(1), 80–96.

Kazancoglu, I., Ozbiltekin-Pala, M., Mangla, S. K., Kumar, A., & Kazancoglu, Y. (2023). Using emerging technologies to improve the sustainability and resilience of supply chains in a fuzzy environment in the context of COVID-19. *Annals of Operations Research*, *322*(1), 217–240.

Kehayov, M., Holder, L., & Koch, V. (2022). Application of artificial intelligence technology in the manufacturing process and purchasing and supply management. *Procedia Computer Science*, *200*, 1209–1217.

Kembro, J., Näslund, D., & Olhager, J. (2017). Information sharing across multiple supply chain tiers: A Delphi study on antecedents. *International Journal of Production Economics*, *193*(June 2016), 77–86. https://doi.org/10.1016/j.ijpe.2017.06.032

Khan, S. A., Agyemang, M., Ishizaka, A., Zaman, S. I., Ali, S. M., & Laval, J. (2021). Barriers and overcoming strategies to multi-tier sustainable supply chain management: an explorative study in an emerging economy. *International Journal of Sustainable Engineering*, *14*(6), 1484–1495.

Khan, S. A., Ali, M. R., & Zaman, I. (2022a). Exploring interrelationships among barriers and enablers of green procurement for a sustainable supply chain. *International Journal of Procurement Management*, *1*(1), 1. https://doi.org/10.1504/ijpm.2022.10052001

Khan, S. A., Hassan, S. M., Kusi-Sarpong, S., Mubarik, M. S., & Fatima, S. (2022b). Designing an integrated decision support system to link supply chain processes performance with time to market. *International Journal of Management Science and Engineering Management*, *17*(1), 66–78.

Khan, S. A., Mubarik, M. S., Kusi-Sarpong, S., Gupta, H., Zaman, S. I., & Mubarik, M. (2022). Blockchain technologies as enablers of supply chain mapping for sustainable supply chains. *Business Strategy and the Environment*.

Khan, M. I., Zaman, S. I., & Khan, S. A. (2023). Relationship and Impact of Block Chain Technology and Supply Chain Management on Inventory Management. In *Blockchain Driven Supply Chain Management* (pp. 53–74). Springer.

Khayyam, S., Ramish, A., Rehman, K., & Syed, A. (2022). Strategizing Radio Frequency Identification (RFID) in the Retail Supply Chains of Pakistan: A Multiple Case Study. *Operations and Supply Chain Management: An International Journal*, *15*(1), 27–40.

Los, U. M. D. E. C. D. E. (2022). Smart business and the social value of AI. *Advanced Series in Managemen.*

Meinlschmidt, J., Schleper, M. C., & Foerstl, K. (2018). Tackling the sustainability iceberg: A transaction cost economics approach to lower tier sustainability management. *International Journal of Operations and Production Management*, *38*(10), 1888–1914. https://doi.org/10.1108/IJOPM-03-2017-0141

Min, H. (2010). Artificial intelligence in supply chain management: Theory and applications. *International Journal of Logistics Research and Applications, 13*(1), 13–39. https://doi.org/10.1080/13675560902736537

Modgil, S., Singh, R. K., & Hannibal, C. (2022). Artificial intelligence for supply chain resilience: learning from Covid-19. *International Journal of Logistics Management, 33*(4), 1246–1268. https://doi.org/10.1108/IJLM-02-2021-0094

Mubarik, M. S., Khan, S. A., Acquaye, A., & Mubarik, M. (2023). Supply chain mapping for improving "visilience": A hybrid multi-criteria decision making based methodology. *Journal of Multi-Criteria Decision Analysis.*

Na, S., Heo, S., Han, S., Shin, Y., & Roh, Y. (2022). Acceptance model of artificial intelligence (AI)-based technologies in construction firms: Applying the Technology Acceptance Model (TAM) in combination with the Technology–Organisation–Environment (TOE) framework. *Buildings, 12*(2), 90.

Naz, F., Agrawal, R., Kumar, A., Gunasekaran, A., Majumdar, A., & Luthra, S. (2022). Reviewing the applications of artificial intelligence in sustainable supply chains: Exploring research propositions for future directions. *Business Strategy and the Environment, 31*(5), 2400–2423.

Nejati, M., Rabiei, S., & Chiappetta Jabbour, C. J. (2017). Envisioning the invisible: Understanding the synergy between green human resource management and green supply chain management in manufacturing firms in Iran in light of the moderating effect of employees' resistance to change. *Journal of Cleaner Production, 168*, 163–172. https://doi.org/10.1016/j.jclepro.2017.08.213

Nozari, H., Szmelter-jarosz, A., & Ghahremani-nahr, J. (2022a). *(AIoT) for the Smart Supply Chain (Case Study :*

Nozari, H., Szmelter-jarosz, A., & Ghahremani-nahr, J. (2022b). Analysis of the challenges of artificial intelligence of things (AIoT) for the smart supply chain (case study: FMCG industries). *Sensors, 22*(8). https://doi.org/10.3390/s22082931

Praveen, U., Farnaz, G., & Hatim, G. (2019). Inventory management and cost reduction of supply chain processes using AI based time-series forecasting and ANN modeling. *Procedia Manufacturing, 38*, 256–263.

Reshad, A. I., Biswas, T., Agarwal, R., Paul, S. K., & Azeem, A. (2023). Evaluating barriers and strategies to sustainable supply chain risk management in the context of an emerging economy. *Business Strategy and the Environment*, (March 2022), 1–20. https://doi.org/10.1002/bse.3367

Rodríguez, L., Giannakis, M., & Cunha, C. (2018). Investigating the Enablers of Big Data Analytics on Sustainable Supply Chain. *Ipsera, Mar*, 1–22.

Sabry Riad Abdel Wanes, O. (2021). Artificial Intelligence Role for Advertising Campaigns development. *International Journal of Artificial Intelligence and Emerging Technology, 4*(1), 1–16.

Sætra, H. S. (2021). Ai in context and the sustainable development goals: Factoring in the unsustainability of the sociotechnical system. *Sustainability (Switzerland), 13*(4), 1–19. https://doi.org/10.3390/su13041738

Sarkis, J., Santibanez Gonzalez, E. D., & Koh, S. C. L. (2019). Effective multi-tier supply chain management for sustainability. *International Journal of Production Economics, 217*, 1–10. https://doi.org/10.1016/j.ijpe.2019.09.014

Shah, H. M., Gardas, B. B., Narwane, V. S., & Mehta, H. S. (2021). The contemporary state of big data analytics and artificial intelligence towards intelligent supply chain risk management: A comprehensive review. *Kybernetes.*

Shankar, S. P., Supriya, M. S., Varadam, D., Kumar, M., Gupta, H., & Saha, R. (2023). A comprehensive study on algorithms and applications of artificial intelligence in diagnosis and prognosis: AI for healthcare. In *Digital Twins and Healthcare: Trends, Techniques, and Challenges* (pp. 35–54). IGI Global.

Su, C.-W., Yuan, X., Umar, M., & Lobonţ, O.-R. (2022). Does technological innovation bring destruction or creation to the labor market? *Technology in Society, 68*, 101905.

Susan, M. (2010). *The Handbook of Technology Management: Supply Chain Management, Marketing and Advertising, and Global Managemen* (1 ed.). New Jersey: John Wiley & Sons, Inc.

Toorajipour, R., Sohrabpour, V., Nazarpour, A., Oghazi, P., & Fischl, M. (2021). Artificial intelligence in supply chain management: A systematic literature review. *Journal of Business Research, 122*(May 2020), 502–517. https://doi.org/10.1016/j.jbusres.2020.09.009

Treiblmaier, H., Mirkovski, K., Lowry, P. B., & Zacharia, Z. G. (2020). The physical internet as a new supply chain paradigm: a systematic literature review and a comprehensive framework. *International Journal of Logistics Management, 31*(2), 239–287. https://doi.org/10.1108/IJLM-11-2018-0284

Vázquez, M. Á., Henarejos, P., Pérez-Neira, A. I., Grechi, E., Voight, A., Gil, J. C., … Lancellotti, R. M. (2020). On the Use of AI for Satellite Communications. *ArXiv, 29*(1), 3–5.

Wang, Y., & Sarkis, J. (2021). Emerging digitalisation technologies in freight transport and logistics: Current trends and future directions. *Transportation Research Part E: Logistics and Transportation Review*, Vol. 148, p. 102291. Elsevier.

Younis, H., Sundarakani, B., & Alsharairi, M. (2022). Applications of artificial intelligence and machine learning within supply chains: systematic review and future research directions. *Journal of Modelling in Management, 17*(3), 916–940.

Zhang, D., Pee, L. G., & Cui, L. (2021). Artificial intelligence in E-commerce fulfillment: A case study of resource orchestration at Alibaba's Smart Warehouse. *International Journal of Information Management, 57*, 102304.

Zhang, G., Yang, Y., & Yang, G. (2023). Smart supply chain management in Industry 4.0: the review, research agenda and strategies in North America. *Annals of Operations Research, 322*(2), 1075–1117. https://doi.org/10.1007/s10479-022-04689-1

Chapter 3
Emerging Disruptive Technologies (EDTs) and Contemporary Supply Chains

Muhammad S. Ahmed, Shiri D. Vivek, and Revansidha D. Chabukswar

3.1 Introduction

Disruptive digital technologies (DTs) have fundamentally transmuted businesses across the globe, redefining how people interact with one another and their surroundings (Büyüközkan and Göçer 2018; Mubarik and Khan 2024). The advent of Industry 4.0 has introduced cutting-edge technologies and methods that significantly alter supply chain and logistics business operations. Robotics and autonomous vehicles, for instance, are rapidly replacing human-powered processes (Akkaya and Kaya 2019). This technological shift necessitates an organizational transformation, emphasizing new methods of value generation and innovative thinking in supply chain management (SCM). Leveraging advancements in the digital framework can save time and effort throughout supply chains (SCs) (Müller and Voigt 2018; Alzarooni et al. 2022), while ensuring products' identification, location, and tracking information are visible at every stage (Kusi-Sarpong et al. 2024).

Specifically, these DTs play a pivotal role in managing traditional SCs by enhancing efficiency, accuracy, and real-time data visibility. Technologies like cloud computing, big data analytics, additive manufacturing, blockchain technology, and the Internet of Things (IoT) are revolutionizing how SC processes operate, enabling faster and more informed decision-making. Contemporary SCs face various challenges and disruptions, including geopolitical conflicts, pandemics, and market

M. S. Ahmed (✉)
Professor of Engineering Management, Eastern Michigan University, Ypsilanti, MI 48085, USA
e-mail: mahmed6@emich.edu

S. D. Vivek
Professor of Marketing and Supply Chain Management, Eastern Michigan University, Ypsilanti, MI 48085, USA

R. D. Chabukswar
Graduate Student Engineering Management, Eastern Michigan University, Ypsilanti, MI 48085, USA

M. S. Mubarik and S. A. Khan (eds.), *Smart Supply Chain Management*,
https://doi.org/10.1007/978-981-96-1333-5_3

uncertainties. DTs offer robust solutions to these challenges, ensuring resilience and adaptability. For instance, the increased data volume in digital supply chains (DSCs) provides new opportunities for automation and improved process management (Junaid et al. 2024). Examples such as robotic warehouses and autonomous freight trucks illustrate the growing impact of automation, which is likely to expand further, necessitating SC specialists to stay abreast of these technological advancements.

Despite extensive research on the role of DTs, it remains crucial to understand how specific technologies like IoT, blockchain, and data science continue to shape SCM (Ahmed and Chabukswar 2023). These technologies not only enhance efficiency and transparency but also drive the digital transformation of corporations, improving overall performance (Ahmed and Chabukswar 2024). The objectives of this chapter are to analyze the impact of emerging disruptive technologies on SCM, explore their current applications, and discuss strategies to leverage these technologies for competitive advantage. The chapter follows a structured sequence, beginning with an overview of the digital transformation in SCs, followed by a comprehensive discussion examining various disruptive technologies and their development over time. It then explores the challenges and opportunities posed by these technologies, concluding with strategic insights for future applications and research.

3.2 Disruptive Technologies in Supply Chains: An Analytical Literature Perspective

Although there is a series of disruptive technologies (DTs), we have selected blockchain technology, additive manufacturing, big data analytics, IoT, and cloud computing. The rationale for selecting these DTs is their recurring appearance in supply chain management (SCM) literature and their anticipated transformative role in SCM. Given the word count and the approach of this chapter, we have adopted a hybrid methodology. In the first section, we review each specific technology, followed by an analysis based on research papers from 2001 to 2022. While acknowledging that most DTs became prominent in scholarly literature post-2009, we have chosen to include the period from 2001 to 2022 to ensure a comprehensive review. This timeframe allows us to capture the initial discussions and foundational studies that paved the way for the widespread adoption and development of these technologies in SCM.

3.2.1 Blockchain Technology

Blockchain quickly rose to the top of the technology stack for usage in financial applications. Blockchain technology is best described as a distributed ledger database for permanently and verifiably recording records of transactions between members.

There has been a lot of revolution and study regarding this distributed technology, which is still in its infancy (Wang et al. 2021). Still, researchers and practitioners have recently focused on other potential uses for Blockchain, like the digital Supply Chain (Mubarik and Shahbaz 2023). Researchers argue that by ensuring data immutability and public access to data streams, blockchain could boost the incoming operations and the efficiency, dependability, and transparency of the entire SC (Ahmed and Chabukswar 2024). In addition to being a new SC network and Internet infrastructure built on distributed applications, blockchain technology also has the potential to change how future business is conducted.

According to claims, blockchain technology may independently address the problem of data veracity and verification without relying on the origin or source of the data. The blockchain application seamlessly integrates into the digital supply chain as a system that links all participating businesses (Susilo and Triana 2018). It has been proven that blockchain increases supply chain visibility. Applications at scale are most likely for goods like infant food, where the final consumer is willing to pay the premium needed to support the technology (Rogerson and Parry 2020). Despite some arguments, blockchain has been emphasized as a powerful, distributed, safe technology in the twenty-first century. It is a digital economic ledger impervious to corruption and can be programmed to record practically any valuable transaction, not just financial ones (Wang et al. 2021). Blockchain is a disruptive technology that can potentially transform sectors on a global scale, add value to businesses and supply chain networks, enhance commerce, and boost the economy. Even though blockchain has garnered notice, very few studies have concentrated on supply chain collaboration and integration (Ahmed and Chabukswar 2024).

3.2.1.1 Adoption of Blockchain Technology in the Smart Supply Chain

In smart supply chains, blockchain technology is a new disruptive technology. Figure 3.1 shows that during the research period from 2001 to 2022, there were several papers on supply chains employing blockchain technology. Most articles were published [C (count) = 596] in 2022, while only three (C = 3) were published in 2017. Beginning in 2019, the number of publications per year increased significantly.

For the study period (2001–2022), 1501 publications were searched in the Web of Science database.

Figure 3.2 indicates the Bibliometrix historiography highlighting the publication of the blockchain technology articles during the study period. During the period 2017–2019, the authors studied the role of blockchain in strengthening cybersecurity and protecting privacy (Kshetri 2017), supply chain theories from the blockchain technology standpoint, information sharing in medication distribution, data securing in additive manufacturing, and supply chain sustainability.

Blockchain technology will help create and test a pharma-co surveillance blockchain system that will facilitate information sharing along the official medication distribution network and is said to be immutable through repeated sequential hashing and fault-tolerant through a consensus method (Sylim et al. 2018). Since

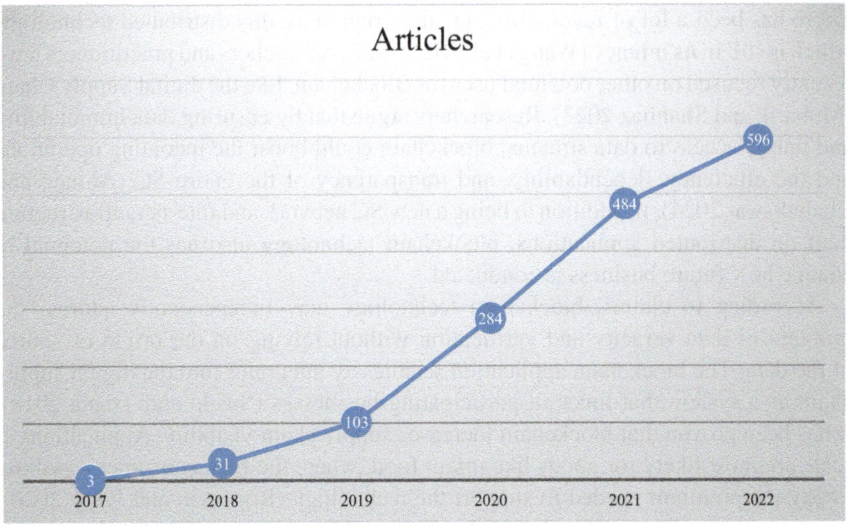

Fig. 3.1 Exploring the evolution of blockchain technology: 2001–2022

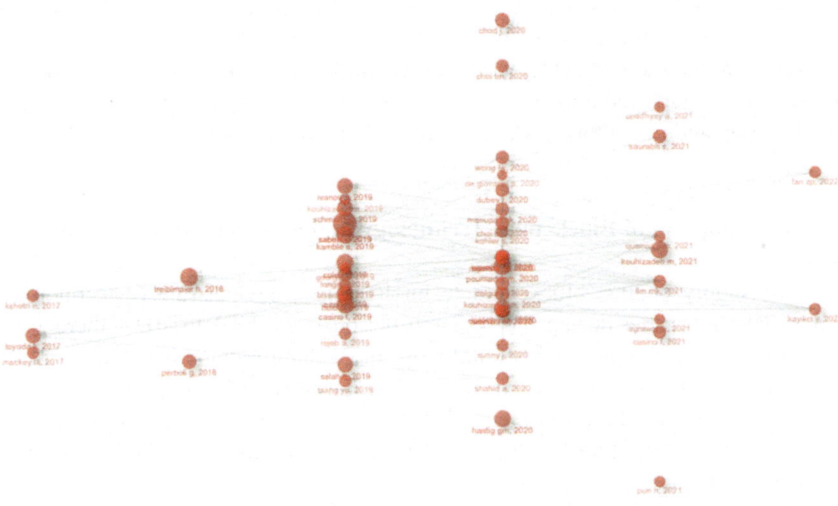

Fig. 3.2 Bibliometrix historiography of blockchain technology: 2001–2022

the majority of the data is presently kept in cloud data centers, it also evaluates how well blockchain functions in comparison to the cloud in terms of security and privacy in several areas. From the perspectives of security and privacy, it illustrates how blockchain-based solutions may be, in many ways, superior to the present IoT ecosystem, which primarily relies on centralized cloud servers provided by service providers (Kshetri 2017). Kshetri (2017) also discussed the importance of Blockchain in determining the root cause of security breaches in IoT device supply chains. The four grand theories are seen from the standpoint of the blockchain's essential characteristics to create theoretical frameworks. Principal agent theory (PAT), transaction cost analysis (TCA), resource-based view (RBV), and network theory (NT) are the major theories (Treiblmaier 2018).

Blockchain technology demonstrates how businesses utilizing blockchain may create secure and interconnected manufacturing infrastructure. It offers a solution for securing and managing the data produced by an end-to-end additive manufacturing process in the aerospace sector (Mandolla et al. 2019). Being a disruptive technology, blockchain can potentially impact sustainable supply chain management. With blockchain technology, small and medium-sized businesses can construct a conceptual model of sustainable supply chain management. Research on sustainable supply chain management is becoming more important as the emphasis on sustainable business processes increases (Nayak and Dhaigude 2019).

During 2020–2022, the authors discuss the application of distributed ledger technology/ blockchain technology in the humanitarian sector, security issues and blockchain solutions for IoT and IIoT, agricultural and food sector, port operations, and logistics management. Although this technology is still in its infancy, donors and government organizations call for developing swift trust and more collaborative relationships among different humanitarian actors.

The Internet of Things (IoT) is becoming increasingly popular, which presents a potential prospect for developing various home automation systems and diverse industrial applications. It helps increase the transparency and traceability of disaster relief materials, information exchanges, and the flow of funds in disaster relief supply chains. Internet of Things (IoT) and Industrial Internet of Things (IIoT) are vulnerable to several cyberattacks, necessitating complex techniques to achieve the needed security and a blockchain-based solution to this threat (Sengupta et al. 2020).

Blockchain technology promises a reliable source of information about the state of farms, inventories, and contracts in agriculture, which helps create trustworthy food supply chains and fosters trust between producers and consumers in an industry where gathering such information is frequently prohibitively expensive. For international trade and transportation services, efficient port logistic operations and management are essential, as are the security and immutability of data stored and transferred during various operational procedures. The all-encompassing solution for such functional operations is provided by blockchain technology (Ahmad et al. 2021).

3.2.2 Additive Manufacturing (3D Printing)

Additive manufacturing (AM) is recognized as being friendly to the environment (Lu and Shi 2022), and metal additive manufacturing (AM) has the potential to be a disruptive technology that profoundly alters the industrial supply chain. Additive manufacturing process parameters should be optimized to maximize build quality and minimize environmental impact and build time (Lu and Shi 2022). According to Stark et al., the most urgent pillar was cybersecurity, and the least critical was additive manufacturing (AM). His findings provide empirical evidence of the effect of the recent Industry 4.0

The traditional supply chain has been significantly altered by additive manufacturing in four essential areas: time, quality, speed, and environmental effect, and in the upcoming years, the industries in India will benefit from this research (Bhasin et al. 2022). Electronic communication technologies are used to transfer design files for additive manufacturing between designer and manufacturer. Material biometrics uses the intrinsic chemical properties of structural materials to generate unique identifiers for authenticating individual products (Masna et al. 2021).

3.2.2.1 Adoption of Additive Manufacturing (AM) in Supply Chain

AM is an emerging innovative technology in digital supply chains. Figure 3.3 shows numerous publications on supply chains using AM technology from 2001 to 2022. 2022 had the most significant number of papers published [C (count) = 107], followed by different years in 2021 (C = 88) and 2020 (C = 80), while it is observed from the graph that 2006, 2009, and 2013 produced the least number of articles C = 1, 1 & 2. The number of publications per year slowly increased and suddenly dropped in 2017 (C = 19). It has been observed that the number of publications has increased in 2018 (C = 32).

AM technology is becoming more prevalent in supply chain management over time. 477 publications were searched in the Web of Science database for the study period (2001–2022). Figure 3.4 indicates the Bibliometrix historiography highlighting the publication of the articles over 20 years with AM technology.

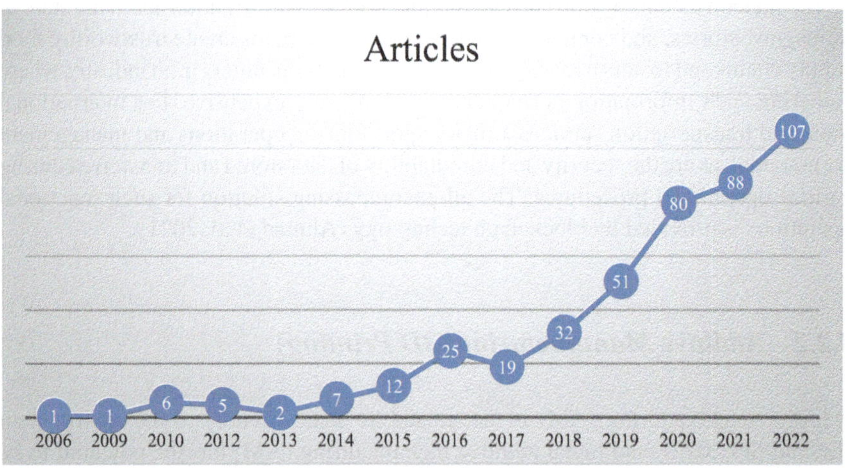

Fig. 3.3 Exploring the evolution of additive manufacturing: 2001–2022

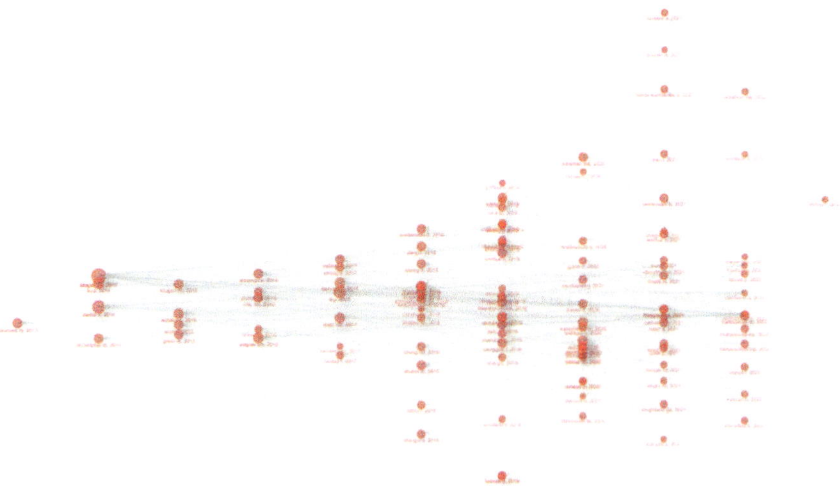

Fig. 3.4 Bibliometrix historiography of additive manufacturing: 2001–2022

Journal articles from 2019 to 2015 examine how network connectivity (average degree) and the number of levels covered in nodes search impact the total production tasks thoroughly performed in an Industry 4.0 production network (Pereira et al. 2018). Chang and Ge (2019) describe a codesign between the flexible electronics supply chain chains to derive practically flexible electronics and sensors for applications. The configuration changes of the reconfigurable limb associated with the three distinct phases of the PMM are analyzed, and the constraints exerted by the reconfigurable limb in various configurations are identified (Ye et al. 2016). The color additives are applied to many food, drug, and cosmetic products, and their safety is critical. Therefore, consistent industry-wide safety standards are needed to ensure quality and safety throughout the supply chain (Simon et al. 2017). The technical solutions identify the practical implications and develop four e-commerce channel designs that reinforce current e-commerce research suggesting that e-commerce channels can lead to both disintermediation and re-intermediation of the Supply Chain.

From 2015 to 2010, the rapid prototyping studies focused on the implementation process of AM. They provided an implementation framework to guide project managers in adopting this new and potentially disruptive technology class (Mellor et al. 2014). New developments are being made, especially in nanoadditives for wood-plastic composites (WPC), including nanocellulose. Recently, the trend of patent registrations for WPCs has shifted to new products or applications (Gardner et al. 2015). With the advent of additive manufacturing, design optimization and the on-demand creation of customized parts have become commonplace manufacturing techniques (Huang et al. 2013).

Agent-based modeling is a well-known approach in supply chain management and simulation. It involves creating computational models where autonomous agents interact with each other and their environment to mimic real-world behaviors and patterns. These models often study complex systems, such as supply chains, where multiple entities interact and influence each other's behavior. One such research demonstrates how an agent-based model can be created to capture the interactions among the various parts of a lubricant additive manufacturing supply chain. Such agent-based modeling is a promising tool for modeling large networks of distributed entities (Behdani et al. 2010).

3.2.3 Big Data Analytics

Big data analytics (BDA) is becoming increasingly significant in every facet of corporate management (Jha et al. 2020). "Velocity, variety, volume, veracity, vision, volatility, verification, validation, variability, and value" are all characteristics of the big data acquired (Abdirad and Krishnan 2020). Data can be recorded in organized, semi-structured, and unstructured formats using various data recording tools. Data from multiple sources, including industrial equipment and systems and organizational and customer management systems, will be routinely gathered and analyzed to enable real-time decision-making (Vaidya et al. 2018).

Organizations need big data and supply chains to handle unstable, dynamic, and global value networks (Kavre et al., n.d.). Recent years have seen a rise in the importance of big data analytics in supply chains as cloud computing enables partners to gather, transmit, store, and process vast amounts of data affordably and exchange the data/information in real-time. Additionally, BDA for supply chains shows considerable promise for process enhancement, cost reduction, and improved supply chain management decision-making (Lee and Mangalaraj 2022. Big data analytics research has produced tools and methods for making data-driven supply chain decisions (Govindan et al. 2018). Big data analytics in supply chain management (SCM) is developing rapidly due to recent developments in machine learning and computer infrastructure.

The global big data market is anticipated to increase from USD 162.6 billion in 2021 to USD 273.4 billion by 2026, at a compound yearly growth rate (CAGR) of 11% from 2021 to 2026, according to marketsandmarkets. com. Big data analytics is predicted to expand at the highest CAGR of all the solutions throughout the projection period and to hold the largest market share (Market sand Markets 2022). Due to climate change, sustainable supply chain management has been an important research topic over the past 20 years. Big data analytics and other recent advancements in information and communication technology, in particular, can assist in building fresh perspectives that can identify supply chain components and individuals whose behaviors are unsustainable and take corrective action (Mageto 2021).

3.2.3.1 Adoption of Big Data (BD) in the Supply Chain

An emerging technology with knowledge and insights acquired from massive amounts of data gives organizations that master this technical innovation a competitive edge (Xu et al. 2021). Figure 3.5 shows that over the study period from 2001 to 2022, BD technology in supply chains was extensively covered in publications, although there were few to none before 2014. Most papers were published in 2022 [C(count) = 108], followed by 2021 (C = 96), and gradually decreased till 2016, with a sudden drop in 2015.

For the study period (2001–2022), 438 publications were searched for in the Web of Science database. The Bibliometrix historiography, which uses big data technology to highlight the publication of papers over 22 years, is shown in Fig. 3.6.

During the research period of 2014–2018, it has been observed that the authors talked about a methodical approach for breaking down big data systems into four sequential modules: data generation, data acquisition, data storage, and data analytics (Hu et al. 2014), and address the problem of data quality in the context of a digital supply chain and suggest techniques for keeping an eye on and managing data quality (Hazen et al. 2014). Dynamic capabilities theory envisions using BDA as a unique information processing capability that gives firms a competitive advantage (Chen et al. 2015). Due to the significant growth in research interest in big data, academics and industry professionals have also attempted to harness its potential for improving operations and supply chain management (Lamba and Singh 2017). Practitioners researched the performance of the sustainable tourism supply chain (STSCP) to determine the significance of big data analytics (BDA) management capabilities, namely BDA planning, BDA investment, BDA coordination, and BDA

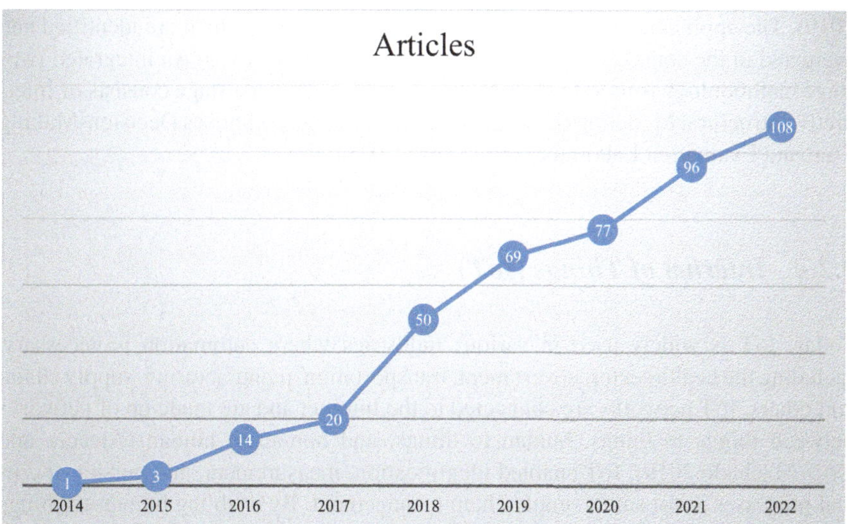

Fig. 3.5 Exploring the evolution of big data analytics: 2001–2022

Fig. 3.6 Bibliometric historiography of big data analytics: 2001–2022

control (Mandal 2018). The authors made an effort to comprehend how organizational capabilities such as big data and predictive analytics (BDPA) might enhance the visibility and coordination of supply chains for humanitarian organizations (Dubey et al. 2018) (Jeble et al. 2020).

During the research period of 2019–2022, supply chain management in disaster assistance is getting new prospects thanks to discussions among practitioners regarding big data analytics. Akter and Wamba (2019) examine how building big data analytics (BDA) skills within an organization can help businesses become more resilient to supply chain disruption events in terms of business risk (Singh and Singh 2019). The application of BDA has considerable obstacles, which are identified and evaluated in the context of Indian industrial supply chains (SC). An integrated two-stage methodology is used to model these obstacles. The first stage consists of Interpretive Structural Modeling (ISM), and the second stage comprises Decision-Making Trial and Evaluation Laboratory (Raut et al. 2021).

3.2.4 Internet of Things (IoT)

Today, IoT is widely used in various industries where automation is necessary, including the health sector, government, transportation, manufacturing, supply chain, and others. IoT networks are connected to the Internet and are made up of networks between things to things, human to things, and human to human (Alcácer and Cruz-Machado 2019). IoT-enabled identification, measurement, and pursuit of critical processes assist smart supply chain management. By enabling human-to-things communication and autonomous coordination among "things" when they are being contained in a facility or being transported between different supply chain organizations, IoT improves supply chain communications (Ben-Daya et al. 2019). In addition, expenditures are decreased, and operations are expedited by eliminating formal procedures (Azizi et al. 2021). Enterprise operations are impacted by IoT innovation, which benefits businesses in the logistics sector (Tu 2018).

The Internet of Things (IoT) or automated handling equipment will be encouraged to replace manual processes in the enterprise management system to handle repetitive jobs in smart supply chain management (Alsudani et al. 2023). Food safety has been elevated to a global priority in recent decades. Despite the issues and difficulties associated with information security and availability, the application of IoT technologies in food traceability demonstrates their usefulness in tracking and recording information about food products along the supply chain (Mehannaoui et al. 2023). The Internet of Things (IoT) has a unique function in supply chain management since it has made it possible to monitor every stage of the chain in real-time to ensure the smooth transit of information and items (Nozari and Nahr 2022). IoT was first used in the supply chain to identify and track products using sensors and handheld devices to manage the movement of information and physical objects (Nozari and Nahr 2022).

3.2.4.1 Adoption of the Internet of Things (IoT) in Supply Chain

The Internet of Things is an emerging disruptive technology in smart supply chains. The Web of Science database was searched for 854 publications across the study period (2001–2022). Figure 3.7 shows that during the research period from 2001 to 2022, there were several papers [C(count) = 397] on supply chains employing the Internet of Things (IoT). Most articles were published (C = 112) in 2021, with a slight reduction in 2022, while only one paper (C = 1) was published in 2012 and 2013. From 2016 through 2020, the number of publications increased gradually; in 2021, there was a considerable increase.

The Bibliometrix historiography, depicted in Fig. 3.8, shows Internet of Things to highlight the publication of papers over 22 years.

In the years 2012 through 2017, the authors discussed information sharing in the supply chain of agricultural products, implications from the development of the Internet of Things (IoT) (Ng et al. 2015), and IoT infrastructure based on radio frequency identification (RFID), ambient intelligence, and multi-agent systems enabling collaborative warehouse order fulfillment (Reaidy et al. 2015). The experiments were carried out to show how the Internet of Things (IoT) may be operationalized in the residential context to capture data on a consumer's consumption of products and the ramifications for reverse supply chains (Parry et al. 2016). During this period, it was also considered how the IoT poses risks to the agricultural Supply Chain, particularly those connected to perception, networks, and application layers (Yan et al. 2017). One of the most effective ways to address the issues with information sharing in the supply chain of agricultural products is through the Internet of Things (IoT) (Yan et al. 2016). An approach for information sharing in the farm

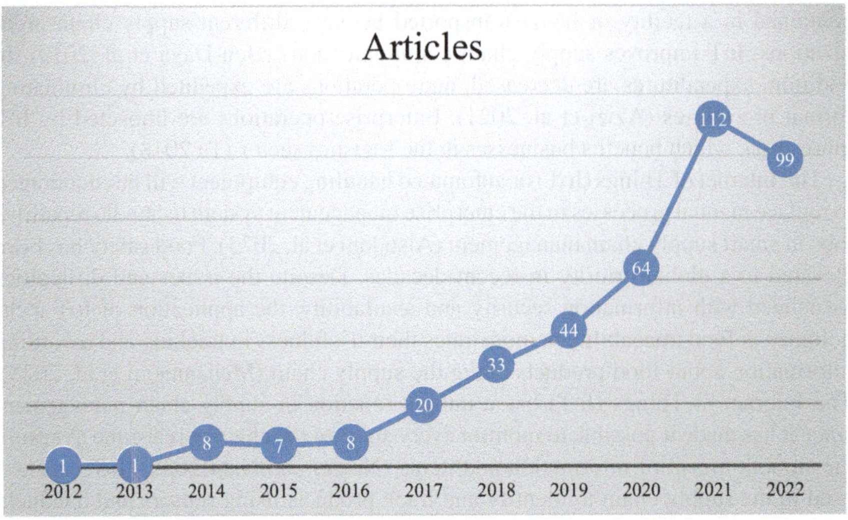

Fig. 3.7 Exploring the evolution of Internet of Things: 2001–2022

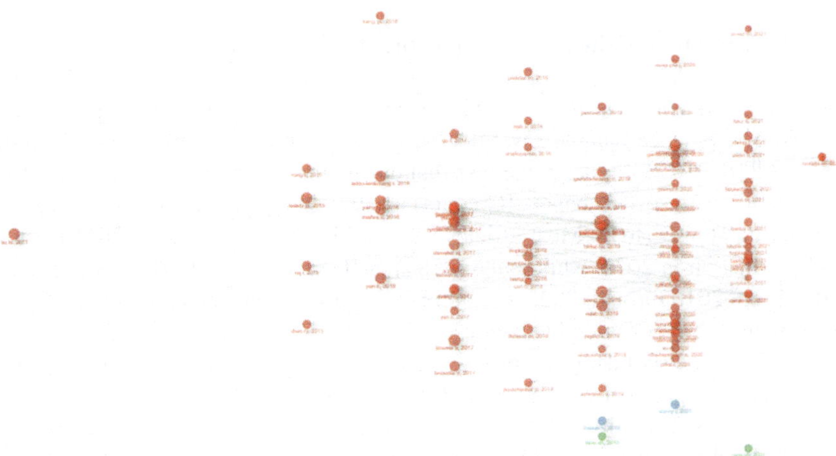

Fig. 3.8 Bibliometric historiography of the Internet of Things: 2001–2022

product supply chain is created, and physical markup language is used to construct and define the information about agricultural items.

Organizations are gradually embracing digital technologies and IoT to improve results. The years 2018 through 2022, the authors and practitioners discuss the crucial role of big data analytics (BDA) and the Internet of Things (IoT) in helping a large logistics company achieve its goals of increasing driver safety, lowering operational costs, and reducing the environmental effect of its fleet of vehicles. The use of an Internet of Things (IoT)-based risk monitoring system (IoTRMS) for regulating occupational safety and product quality hazards in cold chains is crucial for ensuring everyone's personal and occupational safety and the efficiency of the entire cold chain (Tsang et al. 2018). It provides several opportunities to acquire a competitive edge and establishes the standard for future sustainable supply chain processes (Manavalan and Jayakrishna 2019). IoT solution that combines inland vessel navigation with rail scheduling to optimize, manage, and track container transport operations along an intermodal corridor (Munuzuri et al. 2020). A user-friendly open-access fresh products traceability system was developed that focused on interoperability and data exchange (Tagarakis et al. 2021). Such single-chain agriculture supply chain systems have various problems, including storage, scalability optimization, interoperability, security, and privacy. These issues were addressed by Agri-Agriculture Supply Chain Management using blockchain and Internet of Things (SCM-BioT) architecture (Bhat et al. 2022). To secure the supply chain transaction data from privacy and security breaches, a unified solution using distributed ledger technology, i.e., Hyperledger fabric, IoT, and elliptic curve encryption, started showing a sustainable response (Hasan et al. 2022).

3.2.5 Cloud Computing

Recent years have seen the emergence of new IT paradigms, such as cloud computing, which have been used in supply chain management, where the relevance of supply chain efficacy is expanding (Manuel Maqueira et al. 2019). Various supply chains have used cloud-based technologies in many ways. In cloud computing, resources do not reside in specific businesses but in geographically dispersed virtualized and distributed environments that may be accessed on-demand using web-based technologies (Chen and Wu 2013). Large amounts of data are stored in cloud systems from various "business systems, devices, equipment, and sensors on remote servers." Massive volumes of data may be retrieved in real-time with cloud computing platforms (Ghadge et al. 2020). It is necessary to improve data sharing across sites, value chains, departments within a business, and company/organizational boundaries. Suppose enterprises are to successfully create a competitive advantage by effectively allocating resources and implementing competitive practices within the supply chain. In that case, they must pay attention to supply chain relationships and concentrate on collaboration and coordination among their supply chain partners (Lin et al. 2021).

Applications of cloud computing (CC) services for supply chain management (SCM) processes and related enterprise information systems are numerous (Alcácer and Cruz-Machado 2019). Cloud computing (CC) has a lot of advantages because of the decentralized and adaptable structure of global SCs and the rising stability and adaptability of IT operations. For suppliers, SAP, for instance, already offers eight distinct cloud computing services for SC-related IT processes, such as B2B commerce, purchasing, and information exchange (Alcácer and Cruz-Machado

2019) (Shee et al. 2018a). Some studies have compared cloud SCM to more conventional supply chain management methods. They have discovered that cloud SCM gives businesses quicker, more efficient ways to find and use outside technical solutions to boost on-demand services and operational effectiveness (Lin et al. 2021).

3.2.5.1 Adoption of Cloud Computing (CC) in Supply Chain

Throughout the study period (2001–2022), the Web of Science database was searched for a total of 552 publications, although this technique only gained traction after 2010, and no papers were found published before 2010. Figure 3.9 demonstrates that numerous articles [C(count) = 552] on supply chains using cloud computing technology as an emerging disruptive technology in smart supply chains were published throughout the study period of 2001–2022. Only three papers (C = 3) were published in 2010 and 2011, but 70 or more documents (C = 70) were published in 2022. From 2010 to 2021, the number of articles increased gradually before rising quickly in 2022 to double that of the previous year.

In the years 2010 through 2016, the writers talk about how the cloud computing model is being adopted and look at supply chain management from a cloud perspective (Leukel et al. 2011). The Bibliometrix historiography, depicted in Fig. 3.10, shows cloud computing to highlight the publication of papers over 22 years.

Employing organizational information processing theory, one can also determine how a firm's information processing needs and capacities interact to influence its decision to use cloud computing for electronic supply chain management systems (Cegielski et al. 2012). By taking into account the principles of classical diffusion theory as they are articulated within the context of the information processing view, it is possible to analyze the factors that influence a firm's intention to use cloud

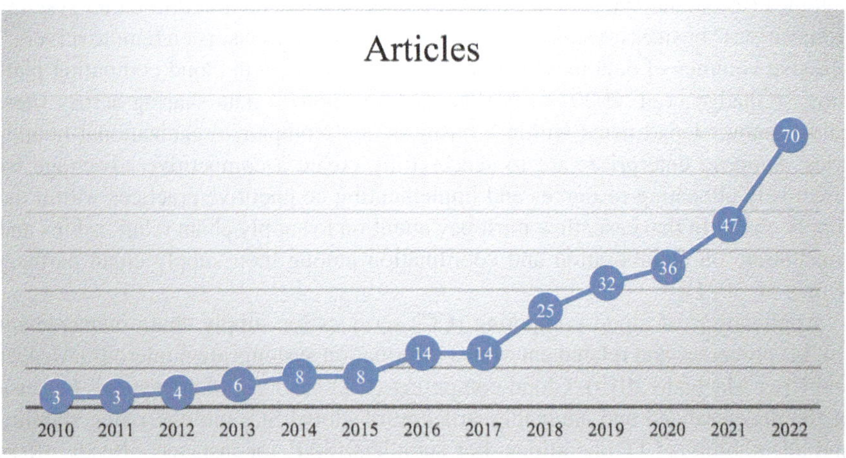

Fig. 3.9 Exploring the evolution of cloud computing: 2001–2022

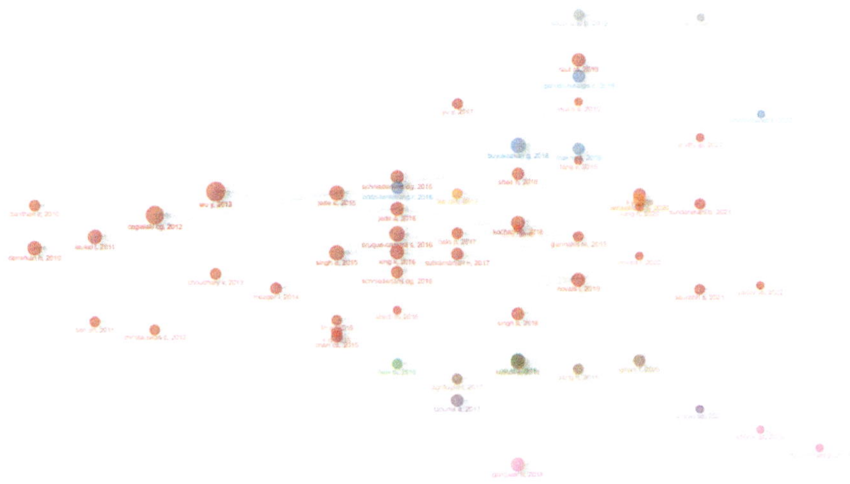

Fig. 3.10 Bibliometrix historiography of cloud computing: 2001–2022

computing technologies in support of its supply chain operations (Wu et al. 2013). Using cloud computing technology (CCT), an integrated system is suggested where all beef supply chain players can reduce and measure carbon emissions at their end with affordable costs and infrastructure (Singh et al. 2015). An analysis of the effects of a particular type of cloud technology (community cloud computing) on a specific kind of supply chain integration has been conducted (Bruque-Camara et al. 2016). Several authors highlighted the use of a cloud-based life-cycle assessment (LCA) platform (Xing et al. 2016; Schniederjans et al. 2016). Such a platform provides dynamic life-cycle data collection and interchange, promotes supply chain collaboration, and is used for environmental footprint assessment (Xing et al. 2016). Cloud computing (CC) affects teamwork and ultimately affects how agile supply chains are for humanitarian purposes (Schniederjans et al. 2016).

From 2017 through 2022, the studies were carried out to evaluate the supply chain risk assessment framework-based cooperative resilience of logistics and cloud computing service providers (Subramanian and Abdulrahman 2017). The authors examined supply chain integration (SCI) as a multidimensional construct in cloud-based technology. They explored how cloud-enabled SCI affects supply chain performance, which would ultimately increase firm sustainability from a resource-based perspective (RBV) (Shee et al. 2018b)—using systems theory and system dynamics to explore cloud computing as an enabler of electronic supply chain management systems (e-SCMs) that improves collaborative information sharing in a multi-echelon hospital supply chain (Kochan et al. 2018). Several methods to improve supply chain responsiveness (SCR) were achieved by developing the architecture of a cloud-based supply chain management (C-SCM) ecosystem (Giannakis et al. 2019). Researchers sought to create a hybrid cloud that primarily aids in integrating the supply chain network with agility and effectiveness (Sundarakani et al. 2021).

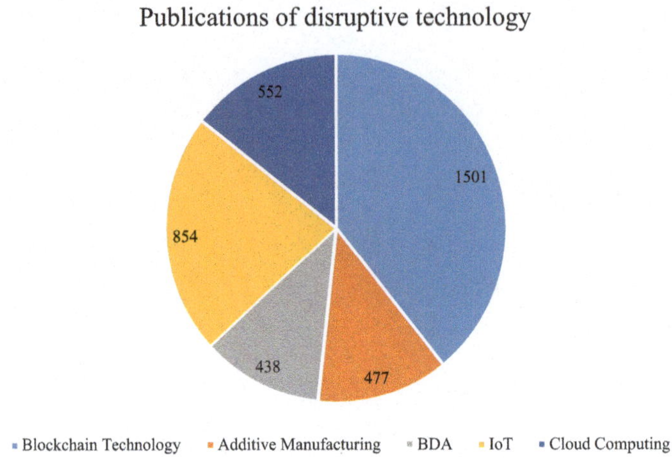

Fig. 3.11 Publications of disruptive technologies: 2001–2022

3.3 Synthesis

Blockchain technology has emerged as the most popular disruptive technology in the smart supply chain management field, according to research on numerous smart technologies between 2001 and 2022, Fig. 3.11. According to a Web of Science search criteria, there have been 1501 articles published on blockchain technology during the research period, compared to 477 for additive manufacturing, 438 for big data analytics, 854 for the Internet of Things, and 552 for cloud computing. The least number of publications have been found for additive manufacturing and big data analytics, indicating a potential research topic for the future.

3.4 Chapter Conclusion

The primary objective of this chapter was to examine the impact of these EDTs on SCM by providing a historical perspective, current applications, and strategic insights for leveraging these technologies. The methodology employed included a thorough review of existing literature, case studies, and empirical data from research papers spanning from 2001 to 2022. The findings reveal that EDTs have significantly enhanced the efficiency and effectiveness of SCM activities. For instance, the adoption of blockchain technology has improved transparency and traceability in supply chains, ensuring data integrity and fostering trust among stakeholders. Additive manufacturing has introduced new possibilities for customization and on-demand production, reducing waste and lead time. Big data analytics has empowered organizations to make data-driven decisions, optimize operations, and predict future trends. IoT has enabled seamless communication and coordination among supply chain entities, enhancing overall operational efficiency. The analysis also highlights the challenges and opportunities associated with implementing these disruptive technologies. One major challenge is the integration of these technologies into existing supply chain infrastructures. Organizations must invest in upgrading their systems and training their workforce to effectively utilize these technologies. Additionally, the rapid pace of technological advancements requires continuous adaptation and innovation. However, the opportunities presented by these technologies far outweigh the challenges. By adopting EDTs, organizations can achieve greater agility, resilience, and sustainability in their supply chains. Blockchain technology, in particular, has emerged as a game-changer in SCM. Its ability to provide an immutable and transparent ledger of transactions has addressed issues of trust and traceability in supply chains. For example, blockchain can be used to track the provenance of products, ensuring that they are ethically sourced and free from counterfeiting. This technology also facilitates efficient and secure data sharing among supply chain partners, reducing the risk of data breaches and enhancing collaboration. The literature review indicates a growing interest in blockchain technology, with a significant increase in publications since 2019, reflecting its rising importance in SCM. Additive manufacturing, or 3D printing, has also shown great promise in transforming supply chains. By enabling localized production and reducing dependence on centralized manufacturing facilities, additive manufacturing can shorten supply chains and reduce transportation costs. This technology allows for rapid prototyping and customization, meeting specific customer demands and reducing time to market. The literature analysis shows a steady increase in research on additive manufacturing, indicating its growing relevance and adoption in various industries. Big data analytics has become a critical tool for SCM, offering valuable insights into supply chain operations and performance. By analyzing large volumes of data from diverse sources, organizations can identify patterns, optimize processes, and make informed decisions. For instance, big data analytics can help predict demand fluctuations, optimize inventory levels, and enhance supplier performance. The literature review reveals a surge in research on big data analytics, particularly in recent years, highlighting its expanding role in SCM. The Internet of Things (IoT) has revolutionized supply chain communication and coordination. IoT devices, such as sensors and RFID tags, provide real-time data on the location, condition, and status of products as they move through the supply chain. This real-time visibility enables organizations to proactively address issues, optimize routes, and improve overall efficiency. The literature analysis indicates a significant increase in IoT-related research, demonstrating its growing importance in modern supply chains. Cloud computing has emerged as a vital enabler of digital transformation in SCM. By providing scalable and cost-effective computing resources, cloud computing allows organizations to store and process vast amounts of data in real-time. This technology facilitates seamless collaboration among supply chain partners, enabling data sharing and integration across different systems. The literature review shows a rising trend in cloud computing research, reflecting its increasing adoption and significance in SCM.

3.5 Future Research Direction

Researchers can focus on developing a unified theory in digital supply chain management by investigating the relevance and integration of pertinent theories from fields such as operations management (e.g., network theory, complexity theory), information systems (e.g., diffusion of innovation, technology acceptance model), and strategic management (e.g., resource-based view, dynamic capabilities). Evaluating their applicability in the distinctive context of digital supply chain management is crucial. It is essential to construct innovative theoretical frameworks that specifically address the intricacies of digital supply chain management, including the impact of enhanced real-time data visibility, the decentralized decision-making capabilities enabled by blockchain technology, and the evolving role of artificial intelligence in supply chain operations. Additionally, it is important to acknowledge that the impact of digital technologies on supply chain management may vary depending on the specific industry, firm size, and geographical context. Therefore, future research could focus on developing theoretical frameworks that incorporate these contextual nuances to provide a more comprehensive understanding of digital supply chain management. Moreover, researchers can shift toward a quantitative research agenda by developing quantitative models and simulations to investigate the effects of specific digital technologies on critical supply chain performance indicators, such as cost optimization, lead time reduction, enhanced agility, and improved sustainability. Comprehensive empirical investigations leveraging quantitative data from actual supply chain environments are needed to test hypotheses, validate theoretical frameworks, and quantify the impact of digital supply chain initiatives. Employing advanced analytical methods, including statistical modeling, data mining, and machine learning, will be essential to extract meaningful insights, forecast future developments, and inform decision-making from the extensive data generated within digital supply chain environments. Pursuing these research avenues can lead to a more comprehensive and nuanced understanding of digital supply chain management, empowering practitioners to fully leverage these transformative technologies and construct more resilient, agile, sustainable, and human-centric supply chain systems in the future.

References

Abdirad, M., & Krishnan, K. (2020). Industry 4.0 in Logistics and Supply Chain Management: A Systematic Literature Review. EMJ Engineering Management Journal, 33(3), 1–15. https://doi.org/10.1080/10429247.2020.1783935

Ahmad, R. W., Hasan, H., Jayaraman, R., Salah, K., & Omar, M. (2021). Blockchain applications and architectures for port operations and logistics management. Research in Transportation Business and Management, 41(October 2020), 100620. https://doi.org/10.1016/j.rtbm.2021.100620

Ahmed, M. S., & Chabukswar, R. D. (2023). DIGITAL SUPPLY CHAIN 2018-2022 – AN ACADEMIC LOOK. World Journal of Engineering Research and Technology, 9(4), 97–109.

Ahmed, M. S., & Chabukswar, R. (2024). DIGITAL SUPPLY CHAIN: A SURVEY OF THE LITERATURE. World Journal of Engineering Research and Technology, 10(5), 56–79.

Akkaya, M., & Kaya, H. (2019). Innovative and smart technologies in logistics. 17th International Logistics and Supply Chain Congress, October 2019, 97–105.

Akter, S., & Wamba, S. F. (2019). Big data and disaster management: a systematic review and agenda for future research. ANNALS OF OPERATIONS RESEARCH, 283(1–2, SI), 939–959. https://doi.org/10.1007/s10479-017-2584-2

Alcácer, V., & Cruz-Machado, V. (2019). Scanning the Industry 4.0: A Literature Review on Technologies for Manufacturing Systems. Engineering Science and Technology, an International Journal, 22(3), 899–919. https://doi.org/10.1016/j.jestch.2019.01.006

Alsudani, M. Q., Jaber, M. M., Ali, M. H., Abd, S. K., Alkhayyat, A., Kareem, Z. H., & Mohhan, A. R. (2023). Smart logistics with IoT-based enterprise management system using global manufacturing. Journal of Combinatorial Optimization, 45(2), 1–31. https://doi.org/10.1007/s10878-022-00977-5

Alzarooni, A. M., Khan, S. A., Gunasekaran, A., & Mubarik, M. S. (2022). Enablers for digital supply chain transformation in the service industry. Annals of Operations Research, 1–25.

Azizi, N., Malekzadeh, H., Akhavan, P., Haass, O., Saremi, S., & Mirjalili, S. (2021). IoT-Blockchain: Harnessing the power of internet of thing and blockchain for smart supply chain. Sensors, 21(18). https://doi.org/10.3390/s21186048

Azmat, M., Ahmed, S., & Mubarik, M. S. (2022). Supply Chain Resilience in the Fourth Industrial Revolution. Supply chain resilience: Insights from theory and practice (pp. 149–163). Cham: Springer International Publishing.

Behdani, B., Lukszo, Z., Adhitya, A., & Srinivasan, R. (2010). Performance analysis of a multi-plant specialty chemical manufacturing enterprise using an agent-based model. COMPUTERS & CHEMICAL ENGINEERING, 34(5, SI), 793–801. https://doi.org/10.1016/j.compchemeng.2010.01.020

Ben-Daya, M., Hassini, E., & Bahroun, Z. (2019). Internet of things and supply chain management: a literature review. International Journal of Production Research, 57(15–16), 4719–4742. https://doi.org/10.1080/00207543.2017.1402140

Bhasin, S., Singari, R. M., Arora, P. K., & Kumar, H. (n.d.). Implications of Additive Manufacturing on Supply Chain Management. JOURNAL OF INDUSTRIAL INTEGRATION AND MANAGEMENT-INNOVATION AND ENTREPRENEURSHIP. https://doi.org/10.1142/S2424862221500299

Bhat, S. A., Huang, N.-F., Sofi, I. B., & Sultan, M. (2022). Agriculture-food supply chain management based on blockchain and IoT: A narrative on enterprise blockchain interoperability. Agriculture-Basel, 12(1). https://doi.org/10.3390/agriculture12010040

Bruque-Camara, S., Moyano-Fuentes, J., & Manuel Maqueira-Marin, J. (2016). Supply chain integration through community cloud: Effects on operational performance. JOURNAL OF PURCHASING AND SUPPLY MANAGEMENT, 22(2), 141–153. https://doi.org/10.1016/j.pursup.2016.04.003

Büyüközkan, G., & Göçer, F. (2018). Digital Supply Chain: Literature review and a proposed framework for future research. Computers in Industry, 97, 157–177. https://doi.org/10.1016/j.compind.2018.02.010

Cegielski, C. G., Jones-Farmer, L. A., Wu, Y., & Hazen, B. T. (2012). Adoption of cloud computing technologies in supply chains An organizational information processing theory approach. INTERNATIONAL JOURNAL OF LOGISTICS MANAGEMENT, 23(2), 184–211. https://doi.org/10.1108/09574091211265350

Chang, J. S., & Ge, T. (2019). A Fully Additive Low-Temperature All-Air Low-Variation Printed/Flexible Electronics with Self-Compensation for Bending: Codesign from Materials, Design, Fabrication, and Applications. PROCEEDINGS OF THE IEEE, 107(10, SI), 2106–2117. https://doi.org/10.1109/JPROC.2019.2941224

Chen, P. yu, & Wu, S. yi. (2013). The impact and implications of on-demand services on market structure. Information Systems Research, 24(3), 750–767. https://doi.org/10.1287/isre.1120.0451

Chen, D. Q., Preston, D. S., & Swink, M. (2015). How the Use of Big Data Analytics Affects Value Creation in Supply Chain Management. JOURNAL OF MANAGEMENT INFORMATION SYSTEMS, 32(4), 4–39. https://doi.org/10.1080/07421222.2015.1138364

Dobrovnik, M., Herold, D. M., Fürst, E., & Kummer, S. (2018). Blockchain for and in Logistics: What to Adopt and Where to Start. Logistics, 2(3), 18. https://doi.org/10.3390/logistics2030018

Dubey, R., Luo, Z., Gunasekaran, A., Akter, S., Hazen, B. T., & Douglas, M. A. (2018). Big data and predictive analytics in humanitarian supply chains: Enabling visibility and coordination in the presence of swift trust. INTERNATIONAL JOURNAL OF LOGISTICS MANAGEMENT, 29(2, SI), 485–512. https://doi.org/10.1108/IJLM-02-2017-0039

Gardner, D. J., Han, Y., & Wang, L. (2015). Wood-Plastic Composite Technology. CURRENT FORESTRY REPORTS, 1(3), 139–150. https://doi.org/10.1007/s40725-015-0016-6

Ghadge, A., Er Kara, M., Moradlou, H., & Goswami, M. (2020). The impact of Industry 4.0 implementation on supply chains. Journal of Manufacturing Technology Management, 31(4), 669–686. https://doi.org/10.1108/JMTM-10-2019-0368

Giannakis, M., Spanaki, K., & Dubey, R. (2019). A cloud-based supply chain management system: effects on supply chain responsiveness. JOURNAL OF ENTERPRISE INFORMATION MANAGEMENT, 32(4), 585–607. https://doi.org/10.1108/JEIM-05-2018-0106

Govindan, K., Cheng, T. C. E., Mishra, N., & Shukla, N. (2018). Big data analytics and application for logistics and supply chain management. Transportation Research Part E: Logistics and Transportation Review, 114(May), 343–349.

Hasan, A. S. M. T., Sabah, S., Haque, R. U., Daria, A., Rasool, A., & Jiang, Q. (2022). Towards convergence of IoT and blockchain for secure supply chain transaction. Symmetry-Basel, 14(1). https://doi.org/10.3390/sym14010064

Hazen, B. T., Boone, C. A., Ezell, J. D., & Jones-Farmer, L. A. (2014). Data quality for data science, predictive analytics, and big data in supply chain management: An introduction to the problem and suggestions for research and applications. INTERNATIONAL JOURNAL OF PRODUCTION ECONOMICS, 154, 72–80. https://doi.org/10.1016/j.ijpe.2014.04.018

Hu, H., Wen, Y., Chua, T.-S., & Li, X. (2014). Toward Scalable Systems for Big Data Analytics: A Technology Tutorial. IEEE ACCESS, 2, 652–687. https://doi.org/10.1109/ACCESS.2014.2332453

Huang, S. H., Liu, P., Mokasdar, A., & Hou, L. (2013). Additive manufacturing and its societal impact: a literature review. INTERNATIONAL JOURNAL OF ADVANCED MANUFACTURING TECHNOLOGY, 67(5–8), 1191–1203. https://doi.org/10.1007/s00170-012-4558-5

Jeble, S., Kumari, S., Venkatesh, V. G., & Singh, M. (2020). Influence of big data and predictive analytics and social capital on performance of humanitarian supply chain Developing framework and future research directions. BENCHMARKING-AN INTERNATIONAL JOURNAL, 27(2), 606–633. https://doi.org/10.1108/BIJ-03-2019-0102

Jha, A. K., Agi, M. A. N., & Ngai, E. W. T. (2020). A note on big data analytics capability development in supply chain. DECISION SUPPORT SYSTEMS, 138. https://doi.org/10.1016/j.dss.2020.113382

Junaid, M., Du, J., Mubarik, M. S., & Shahzad, F. (2024). Creating a sustainable future through Industry 4.0 technologies: Untying the role of circular economy practices and supply chain visibility. Business Strategy and the Environment.

Kavre, M., Narwane, V. S., Gardas, B. B., & Sunnapwar, V. (n.d.). Role of human factors in cloud manufacturing adoption across manufacturing micro, small and medium enterprises. International Journal of Computer Integrated Manufacturing. https://doi.org/10.1080/0951192X.2022.2128215

Khan, S. A., Mubarik, M. S., Kusi-Sarpong, S., Gupta, H., Zaman, S. I., & Mubarik, M. (2022). Blockchain technologies as enablers of supply chain mapping for sustainable supply chains. Business strategy and the environment, 31(8), 3742–3756.

Khan, S. A., Gupta, H., Gunasekaran, A., Mubarik, M. S., & Lawal, J. (2023). A hybrid multi-criteria decision-making approach to evaluate interrelationships and impacts of supply chain performance factors on pharmaceutical industry. Journal of Multi-Criteria Decision Analysis, 30(1–2), 62–90.

Kochan, C. G., Nowicki, D. R., Sauser, B., & Randall, W. S. (2018). Impact of cloud-based information sharing on hospital supply chain performance: A system dynamics framework. INTERNATIONAL JOURNAL OF PRODUCTION ECONOMICS, 195, 168–185. https://doi.org/10.1016/j.ijpe.2017.10.008

Kshetri, N. (2017). Blockchain's roles in strengthening cybersecurity and protecting privacy. TELECOMMUNICATIONS POLICY, 41(10, SI), 1027–1038. https://doi.org/10.1016/j.telpol.2017.09.003

Kusi-Sarpong, S., Mubarik, M. S., Khan, S. A., Brown, S., & Mubarak, M. F. (2022). Intellectual capital, blockchain-driven supply chain and sustainable production: Role of supply chain mapping. Technological Forecasting and Social Change, 175, 121331.

Kusi-Sarpong, S., Mubarik, M. S., & Khan, S. A. (2024). Supply chain mapping for "visilience": Role of blockchain-driven supply chain management. In The Palgrave Handbook of Supply Chain Management (pp. 585–599). Cham: Springer International Publishing.

Lamba, K., & Singh, S. P. (2017). Big data in operations and supply chain management: current trends and future perspectives. PRODUCTION PLANNING & CONTROL, 28(11–12, SI), 877–890. https://doi.org/10.1080/09537287.2017.1336787

Lee, I., & Mangalaraj, G. (2022). Big data analytics in supply chain management: A systematic literature review and research directions. Big Data and Cognitive Computing, 6(1). https://doi.org/10.3390/bdcc6010017

Leukel, J., Kirn, S., & Schlegel, T. (2011). Supply Chain as a Service: A Cloud Perspective on Supply Chain Systems. IEEE SYSTEMS JOURNAL, 5(1), 16–27. https://doi.org/10.1109/JSYST.2010.2100197

Lin, M., Lin, C., & Chang, Y. S. (2021). The impact of using a cloud supply chain on organizational performance. Journal of Business and Industrial Marketing, 36(1), 97–110. https://doi.org/10.1108/JBIM-04-2019-0154

Lu, C., & Shi, J. (2022). Simultaneous consideration of relative density, energy consumption, and build time for selective laser melting of Inconel 718: A multi-objective optimization study on process parameter selection. JOURNAL OF CLEANER PRODUCTION, 369. https://doi.org/10.1016/j.jclepro.2022.133284

Mageto, J. (2021). Big data analytics in sustainable supply chain management: A focus on manufacturing supply chains. Sustainability, 13(13). https://doi.org/10.3390/su13137101

Manavalan, E., & Jayakrishna, K. (2019). A review of Internet of Things (IoT) embedded sustainable supply chain for industry 4.0 requirements. COMPUTERS & INDUSTRIAL ENGINEERING, 127, 925–953. https://doi.org/10.1016/j.cie.2018.11.030

Mandal, S. (2018). Exploring the influence of big data analytics management capabilities on sustainable tourism supply chain performance: the moderating role of technology orientation. JOURNAL OF TRAVEL & TOURISM MARKETING, 35(8), 1104–1118. https://doi.org/10.1080/10548408.2018.1476302

Mandolla, C., Petruzzelli, A. M., Percoco, G., & Urbinati, A. (2019). Building a digital twin for additive manufacturing through the exploitation of blockchain: A case analysis of the aircraft industry. COMPUTERS IN INDUSTRY, 109, 134–152. https://doi.org/10.1016/j.compind.2019.04.011

Manuel Maqueira, J., Moyano-Fuentes, J., & Bruque, S. (2019). Drivers and consequences of an innovative technology assimilation in the supply chain: cloud computing and supply chain integration. International Journal of Production Research, 57(7), 2083–2103. https://doi.org/10.1080/00207543.2018.1530473

Market sand Markets. (2022). Big Data Market by Solutions & Services. MarketsandMarkets. Available Online: https://Www.Marketsandmarkets.Com/Market-Reports/Big-Data-Market-1068.Html. (Accessed).

Masna, N. V. R., Huan, J., Mandal, S., & Bhunia, S. (2021). NQR sensitive embedded signatures for authenticating additively manufactured objects. Scientific Reports, 11(1). https://doi.org/10.1038/s41598-021-91531-6

Mehannaoui, R., Mouss, K. N., & Aksa, K. (2023). IoT-based food traceability system: Architecture, technologies, applications, and future trends. Food Control, 145(September 2022), 109409. https://doi.org/10.1016/j.foodcont.2022.109409

Mellor, S., Hao, L., & Zhang, D. (2014). Additive manufacturing: A framework for implementation. INTERNATIONAL JOURNAL OF PRODUCTION ECONOMICS, 149(SI), 194–201. https://doi.org/10.1016/j.ijpe.2013.07.008

Mubarik, M. S., & Khan, S. A. (2024). The Theory, Methods and Application of Managing Digital Supply Chains. Emerald Publishing Limited: Leeds. https://doi.org/10.1002/bse.3777

Mubarik, M. S., & Shahbaz, M. (2023). Blockchain Driven Supply Chain Management. Springer.

Müller, J. M., & Voigt, K. I. (2018). The Impact of Industry 4.0 on Supply Chains in Engineer-to-Order Industries—An Exploratory Case Study. IFAC-PapersOnLine, 51(11), 122–127. https://doi.org/10.1016/j.ifacol.2018.08.245

Munuzuri, J., Onieva, L., Cortes, P., & Guadix, J. (2020). Using IoT data and applications to improve port-based intermodal supply chains. Computers & Industrial Engineering, 139. https://doi.org/10.1016/j.cie.2019.01.042

Nayak, G., & Dhaigude, A. S. (2019). A conceptual model of sustainable supply chain management in small and medium enterprises using blockchain technology. Cogent Economics and Finance, 7(1). https://doi.org/10.1080/23322039.2019.1667184

Ng, I., Scharf, K., Pogrebna, G., & Maull, R. (2015). Contextual variety, Internet-of-Things and the choice of tailoring over platform: Mass customisation strategy in supply chain management. INTERNATIONAL JOURNAL OF PRODUCTION ECONOMICS, 159(SI), 76–87. https://doi.org/10.1016/j.ijpe.2014.09.007

Nozari, H., & Nahr, J. G. (2022). The Impact of Blockchain Technology and The Internet of Things on the Agile and Sustainable Supply Chain. International Journal of Innovation in Engineering, 2(2), 33–41. https://ijie.ir/index.php/ijie/article/view/67

Parry, G. C., Brax, S. A., Maull, R. S., & Ng, I. C. L. (2016). Operationalising IoT for reverse supply: the development of use-visibility measures. SUPPLY CHAIN MANAGEMENT-AN INTERNATIONAL JOURNAL, 21(2), 228–244. https://doi.org/10.1108/SCM-10-2015-0386

Pereira, A., Simonetto, E. de O., Putnik, G., & de Castro, H. C. (2018). HOW CONNECTIVITY AND SEARCH FOR PRODUCERS IMPACT PRODUCTION IN INDUSTRY 4.0 NETWORKS. BRAZILIAN JOURNAL OF OPERATIONS & PRODUCTION MANAGEMENT, 15(4), 528–534. https://doi.org/10.14488/BJOPM.2018.v15.n4.a6

Piprani, A. Z., Jaafar, N. I., Ali, S. M., Mubarik, M. S., & Shahbaz, M. (2022). Multi-dimensional supply chain flexibility and supply chain resilience: the role of supply chain risks exposure. Operations Management Research, 15(1), 307–325.

Qader, G., Junaid, M., Abbas, Q., & Mubarik, M. S. (2022). Industry 4.0 enables supply chain resilience and supply chain performance. Technological Forecasting and Social Change, 185, 122026.

Raut, R. D., Yadav, V. S., Cheikhrouhou, N., Narwane, V. S., & Narkhede, B. E. (2021). Big data analytics: Implementation challenges in Indian manufacturing supply chains. COMPUTERS IN INDUSTRY, 125. https://doi.org/10.1016/j.compind.2020.103368

Reaidy, P. J., Gunasekaran, A., & Spalanzani, A. (2015). Bottom-up approach based on Internet of Things for order fulfillment in a collaborative warehousing environment. INTERNATIONAL JOURNAL OF PRODUCTION ECONOMICS, 159(SI), 29–40. https://doi.org/10.1016/j.ijpe.2014.02.017

Rogerson, M., & Parry, G. C. (2020). Blockchain: case studies in food supply chain visibility. Supply Chain Management, 25(5), 601–614. https://doi.org/10.1108/SCM-08-2019-0300

Schniederjans, D. G., Ozpolat, K., & Chen, Y. (2016). Humanitarian supply chain use of cloud computing. SUPPLY CHAIN MANAGEMENT-AN INTERNATIONAL JOURNAL, 21(5), 569–588. https://doi.org/10.1108/SCM-01-2016-0024

Sengupta, J., Ruj, S., & Das Bit, S. (2020). A Comprehensive Survey on Attacks, Security Issues and Blockchain Solutions for IoT and IIoT. Journal of Network and Computer Applications, 149(October 2019), 102481. https://doi.org/10.1016/j.jnca.2019.102481

Shee, H., Miah, S. J., Fairfield, L., & Pujawan, N. (2018a). The impact of cloud-enabled process integration on supply chain performance and firm sustainability: the moderating role of top management. Supply Chain Management, 23(6), 500–517. https://doi.org/10.1108/SCM-09-2017-0309

Shee, H., Miah, S. J., Fairfield, L., & Pujawan, N. (2018b). The impact of cloud-enabled process integration on supply chain performance and firm sustainability: the moderating role of top management. Supply Chain Management-An International Journal, 23(6), 500–517. https://doi.org/10.1108/SCM-09-2017-0309

Simon, J. E., Decker, E. A., Ferruzzi, M. G., Giusti, M. M., Mejia, C. D., Goldschmidt, M., & Talcott, S. T. (2017). Establishing Standards on Colors from Natural Sources. JOURNAL OF FOOD SCIENCE, 82(11), 2539–2553. https://doi.org/10.1111/1750-3841.13927

Singh, N. P., & Singh, S. (2019). Building supply chain risk resilience Role of big data analytics in supply chain disruption mitigation. BENCHMARKING-AN INTERNATIONAL JOURNAL, 26(7), 2318–2342. https://doi.org/10.1108/BIJ-10-2018-0346

Singh, A., Mishra, N., Ali, S. I., Shukla, N., & Shankar, R. (2015). Cloud computing technology: Reducing carbon footprint in beef supply chain. INTERNATIONAL JOURNAL OF PRODUCTION ECONOMICS, 164, 462–471. https://doi.org/10.1016/j.ijpe.2014.09.019

Singh, S., Kumar, R., Panchal, R., & Tiwari, M. K. (2021). Impact of COVID-19 on logistics systems and disruptions in food supply chain. International journal of production research, 59(7), 1993–2008.

Subramanian, N., & Abdulrahman, M. D. (2017). Logistics and cloud computing service providers' cooperation: a resilience perspective. PRODUCTION PLANNING & CONTROL, 28(11–12, SI), 919–928. https://doi.org/10.1080/09537287.2017.1336793

Sundarakani, B., Kamran, R., Maheshwari, P., & Jain, V. (2021). Designing a hybrid cloud for a supply chain network of Industry 4.0: a theoretical framework. BENCHMARKING-AN INTERNATIONAL JOURNAL, 28(5, SI), 1524–1542. https://doi.org/10.1108/BIJ-04-2018-0109

Susilo, F. A., & Triana, Y. S. (2018). Digital supply chain development in blockchain technology using Rijndael algorithm 256. IOP Conference Series: Materials Science and Engineering, 453(1), 1–12. https://doi.org/10.1088/1757-899X/453/1/012075

Sylim, P., Liu, F., Marcelo, A., & Fontelo, P. (2018). Blockchain Technology for Detecting Falsified and Substandard Drugs in Distribution: Pharmaceutical Supply Chain Intervention. JMIR Research Protocols, 7(9). https://doi.org/10.2196/10163

Tagarakis, A. C., Benos, L., Kateris, D., Tsotsolas, N., & Bochtis, D. (2021). Bridging the Gaps in Traceability Systems for Fresh Produce Supply Chains: Overview and Development of an Integrated IoT-Based System. Applied Sciences-Basel, 11(16). https://doi.org/10.3390/app11167596

Treiblmaier, H. (2018). The impact of the blockchain on the supply chain: a theory-based research framework and a call for action. SUPPLY CHAIN MANAGEMENT-AN INTERNATIONAL JOURNAL, 23(6), 545–559. https://doi.org/10.1108/SCM-01-2018-0029

Tsang, Y. P., Choy, K. L., Wu, C. H., Ho, G. T. S., Lam, C. H. Y., & Koo, P. S. (2018). An Internet of Things (IoT)-based risk monitoring system for managing cold supply chain risks. INDUSTRIAL MANAGEMENT & DATA SYSTEMS, 118(7, SI), 1432–1462. https://doi.org/10.1108/IMDS-09-2017-0384

Tu, M. (2018). An exploratory study of internet of things (IoT) adoption intention in logistics and supply chain management a mixed research approach. International Journal of Logistics Management, 29(1), 131–151. https://doi.org/10.1108/IJLM-11-2016-0274

Vaidya, S., Ambad, P., & Bhosle, S. (2018). Industry 4.0—A Glimpse. Procedia Manufacturing, 20, 233–238. https://doi.org/10.1016/j.promfg.2018.02.034

Wang, M., Wu, Y., Chen, B., & Evans, M. (2021). Blockchain and Supply Chain Management: A New Paradigm for Supply Chain Integration and Collaboration. OPERATIONS AND SUPPLY CHAIN MANAGEMENT-AN INTERNATIONAL JOURNAL, 14(1), 111–122. https://doi.org/10.31387/oscm0440290

Wu, Y., Cegielski, C. G., Hazen, B. T., & Hall, D. J. (2013). Cloud Computing in Support of Supply Chain Information System Infrastructure: Understanding When to go to the Cloud. JOURNAL OF SUPPLY CHAIN MANAGEMENT, 49(3), 25–41. https://doi.org/10.1111/j.1745-493x.2012.03287.x

Xing, K., Qian, W., & Zaman, A. U. (2016). Development of a cloud-based platform for footprint assessment in green supply chain management. JOURNAL OF CLEANER PRODUCTION, 139, 191–203. https://doi.org/10.1016/j.jclepro.2016.08.042

Xu, M., Cui, Y., Hu, M., Xu, X., Zhang, Z., Liang, S., & Qu, S. (2019). Supply chain sustainability risk and assessment. Journal of Cleaner Production, 225, 857–867.

Xu, J., Pero, M. E. P., Ciccullo, F., & Sianesi, A. (2021). On relating big data analytics to supply chain planning: towards a research agenda. INTERNATIONAL JOURNAL OF PHYSICAL DISTRIBUTION & LOGISTICS MANAGEMENT, 51(6, SI), 656–682. https://doi.org/10.1108/IJPDLM-04-2020-0129

Yan, B., Yan, C., Ke, C., & Tan, X. (2016). Information sharing in supply chain of agricultural products based on the Internet of Things. INDUSTRIAL MANAGEMENT & DATA SYSTEMS, 116(7), 1397–1416. https://doi.org/10.1108/IMDS-12-2015-0512

Yan, B., Wang, X., & Shi, P. (2017). Risk assessment and control of agricultural supply chains under Internet of Things. AGREKON, 56(1), 1–12. https://doi.org/10.1080/03031853.2017.1284680

Ye, W., Fang, Y., Zhang, K., & Guo, S. (2016). Mobility variation of a family of metamorphic parallel mechanisms with reconfigurable hybrid limbs. ROBOTICS AND COMPUTER-INTEGRATED MANUFACTURING, 41, 145–162. https://doi.org/10.1016/j.rcim.2016.03.006

Chapter 4
Capitalizing Innovation Capabilities for Smart Value Chain Management in Collaborative Innovation Networks

Muhammad Faraz Mubarak, Monika Petraite, Mubarra Shabbir, Richard Evans, and Anna Pilkova

4.1 Introduction

Technological environments have become increasingly dynamic, presenting firms with both new opportunities and significant challenges (Mohrman & von Glinow 1990). In this context, sudden, unpredictable, and continuous changes, such as technological discontinuities and shortened innovation life cycles, are altering the value of existing technologies (Amis 2018). Consequently, many firms are focusing their efforts on technology upgrading to maintain competitiveness (Radosevic & Yoruk 2018). This technology upgrading (TU) has the potential to transform entire value chains and supply chains (Ghobakhloo et al. 2023a, b). The implementation of smart technologies, characteristic of Industry 4.0, is crucial in accelerating digital transformation and making supply chain networks smart and agile (Kusi-Sarpong et al. 2022). However, achieving smart supply chain and value chain networks requires technology upgrading, which fundamentally relies on the quality of prior Technological Learning (TL) (Petraite et al. 2022). Technology upgrading is critical because it enhances a firm's ability to stay competitive by advancing its technological capabilities. This process begins with TL, where firms acquire, assimilate, and apply knowledge of new and smart technologies to their operations. Through TL, firms can adapt and

---

M. F. Mubarak (✉) · R. Evans
Faculty of Computer Science, Dalhousie University, Halifax, NS, Canada
e-mail: m.mubarak@dal.ca

M. Petraite · M. Shabbir
School of Economics and Business, Kaunas University of Technology, Kaunas, Lithuania

A. Pilkova
Faculty of Management, Comenius University Bratislava, Bratislava, Slovakia

M. F. Mubarak
Faculty of Economics and Management, Vytautas Magnus University, Kaunas, Lithuania

M. S. Mubarik and S. A. Khan (eds.), *Smart Supply Chain Management*,
https://doi.org/10.1007/978-981-96-1333-5_4

thrive in a rapidly changing environment, thereby creating a competitive edge. Participation in collaborative innovation networks, such as open innovation communities, supply chain networks, and global value chains (GVCs), offers firms opportunities to learn from and share experiences with others, further enhancing their TL. This study adopts a linear approach to the smartization of networks, including value chain networks, supply chain networks, and other innovation networks. This approach starts with technological learning, which leads to technology upgrading and eventually culminates in the creation of smart networks for supply chains, value chains, and collaborative innovation (Ghobakhloo et al. 2023a, b). The linear approach can be summarized as:

Technological Learning → Technology Upgrading → Smartization of Networks

In this context, firms can capitalize on technology upgrading by adopting a continuous learning approach (Kusi-Sarpong et al. 2022). Despite the importance of technology upgrading, the mechanisms to enable it, particularly through different categories of learning, remain largely unaddressed (Petraite et al. 2022). Existing research indicates that technological learning enables firms to detect technological shifts that threaten existing operations and provide potential alternatives (Dodgson et al. 2022). Shabbir et al. (2024) and Petraite et al. (2022) found that technology-based firms can enhance their technological learning by collaborating with partners in collaborative innovation networks. Therefore, to remain competitive, firms should enhance technological learning by participating in such networks, thereby preparing for technology upgrading. Innovation capabilities play a vital role in improving firms' technological learning, although there are inconsistencies regarding the type of capabilities and their relevance (El-Awad et al. 2017). In the context of technology upgrading, researchers have primarily focused on technical capabilities, such as those related to the technological side of new product development (NPD) (Rubera et al. 2016; Qin et al. 2021). However, this study argues that technological learning is strongly associated with the sharing of tacit knowledge and other interactions between network partners. Therefore, a social and relational view of innovation capabilities is critical for improving technological learning, which can drive successful technology upgrading (Ghobakhloo et al. 2023a, b). In this context, collaborative innovation networks not only facilitate the development of technological learning but also enable firms to upgrade their capabilities (Radosevic & Yoruk 2018). These networks provide a platform for firms to engage in knowledge-based interactions, fostering an environment of trust, social capital, innovation culture, and networking capabilities. These elements are essential for effective interactions that enhance technological learning and, subsequently, technology upgrading. Against this backdrop, this study aims to explore the role of the interactive side of innovation capabilities within collaborative innovation networks, with the goal of achieving technology upgrading through technological learning. Also, we explore the relationships and interactions between innovation capabilities toward technologically advance the networks (e.g., smart supply chain, open innovation, and global value chains). This research is critically important as it contributes to the development of smarter and more agile networks. By emphasizing the relational view of innovation capabilities, this study offers new insights into how firms can leverage collaborative innovation networks to drive technology upgrading, thereby enhancing their competitiveness in a dynamic technological environment.

4.2 Technological Learning and Technology Upgrading

Technological learning is a dynamic and ongoing process whereby firms acquire, assimilate, and apply technological knowledge to their current operations. This process is essential for driving technology-related innovation (Oqubay & Ohno 2019) and technological accumulation, which refers to the gradual development of knowledge about new technologies. This accumulation enables firms to adapt and thrive in rapidly changing technological environments (Dodgson 2018). Additionally, researchers posit that technological learning plays a critical role in establishing a knowledge foundation on which innovation can be achieved (Zahra et al. 2000). According to Dodgson (1991, p. 6), technological learning (TL) is defined as "the ways firms build and supplement their knowledge bases on technologies, products, and processes and develop and improve the use of the broad skills of their workforce.“ The degree of novelty achieved through technological learning is contingent upon various factors, including the specific context and individuals involved in the process. The capacity of firms to integrate organization-specific technologies and cultivate technological skills relies significantly on the level and quality of technological learning (Zahra et al. 2000). By enhancing their technological learning capabilities, firms can more easily adapt to technological shifts and address evolving market demands. This adaptability positions firms to successfully undertake technology upgrading initiatives (Kruss et al. 2015).

Technology upgrading refers to the process of enhancing and advancing technological capabilities within a firm (Ghobakhloo et al. 2023a, b). This involves removing obsolete technologies, upgrading and improving existing ones, and leveraging technological advancements to drive innovation and gain a competitive edge (Radosevic & Yoruk 2018). Technological learning serves as a vital enabler of technology upgrading, equipping firms with the knowledge and skills required to adapt quickly to changing technological environments (Petraite et al. 2022). Moreover, technological learning impacts firm innovation at multiple levels. At the instrumental level, technological learning enhances firms' ability to use technology to achieve specific objectives and performance outcomes (Ghobakhloo et al. 2023a, b). It enables firms to effectively apply technological knowledge to improve processes, increase efficiency, and enhance productivity (Dodgson et al. 2022). At the innovative level, technological learning fosters the development of novel products and services. By continuously learning and integrating new technological knowledge, firms can create innovative offerings that meet evolving customer needs and preferences (Petraite et al. 2022). Furthermore, at the creative level, technological learning fuels breakthrough innovations and transformative changes. It facilitates the exploration of new technological frontiers, the generation of disruptive ideas, and the pursuit of radical innovations that can reshape industries and markets (Zahra et al. 2000). Firms that prioritize and invest in enhancing their technological learning capabilities are better positioned to undertake technology upgrading initiatives. By actively engaging in technological learning processes, firms can acquire the knowledge, skills, and insights needed to identify emerging technologies, assess their potential impact, and effectively integrate them into their operations. Ultimately, technological learning serves as a critical driver for technology upgrading (Radosevic & Yoruk 2018). By continuously acquiring and applying new technological knowledge, firms can adapt, innovate, and upgrade their technological capabilities to remain competitive and thrive (Mubarak et al. 2023a, b).

Furthermore, technology upgrading is intricately linked with the advancement of innovation networks, notably supply chain networks. The process of technology upgrading involves not only the enhancement of a firm's internal technological capabilities but also the improvement of its interactions with external partners within the supply chain. By improving smart supply capabilities and digitally transforming entire global value chains, firms can significantly enhance their operational efficiency and competitiveness. Digital transformation, which encompasses the integration of advanced technologies such as artificial intelligence (AI), the Internet of Things (IoT), and big data analytics, plays a crucial role in this context (Ghobakhloo et al. 2023a, b). These technologies enable firms to optimize supply chain operations by providing real-time data and insights, enhancing decision-making processes, and facilitating more effective coordination among network partners. The implementation of AI in supply chains allows for predictive analytics, which can foresee potential disruptions and suggest optimal responses. IoT devices, on the other hand, can track and monitor goods in real-time, ensuring better inventory management and reducing the risk of stockouts or overstock situations. Big data analytics helps in analyzing vast amounts of data generated across the supply chain to identify patterns and trends that can be leveraged for strategic decision-making (Ghobakhloo et al. 2023a, b). This holistic approach to supply chain management not only improves the efficiency and responsiveness of supply chains but also enhances their resilience to disruptions.

Moreover, digital transformation facilitates the integration of new technologies and innovative practices across the entire value chain. This integration is pivotal for fostering innovation networks that are agile, responsive, and capable of sustaining competitive advantages in a rapidly evolving market landscape (Radosevic & Yoruk 2018). For instance, blockchain technology can be used to ensure transparency and traceability in the supply chain, enhancing trust among partners and improving the overall reliability of the network (Ghobakhloo et al., 2024). Additionally, advanced robotics and automation can streamline manufacturing processes, reducing lead times and increasing production efficiency (Ghobakhloo et al. 2023a, b). The transformative impact of digital technologies extends beyond operational improvements. It enables firms to engage in collaborative innovation with their supply chain partners, fostering a culture of continuous improvement and co-creation. This collaborative approach can lead to the development of new products and services that are better aligned with market demands and customer expectations (Kusi-Sarpong et al. 2022). Furthermore, by leveraging digital platforms, firms can create more interconnected and interdependent supply chain networks that are capable of responding swiftly to changes in the external environment.

4.3 Innovation Capabilities

Trust

Trust is a critical factor in accelerating and facilitating the advancement of networks, such as smart supply chain networks, global value chains, and open innovation networks. It plays a pivotal role in enabling collaborative learning within these innovation networks (Doyle, 2023; Petraite et al. 2022; Mubarak & Petraite 2020). Such networks rely heavily on effective knowledge and information sharing among partners, which is significantly enhanced by the presence of trust. In open innovation communities, where multiple firms, suppliers, customers, and other stakeholders participate, trust becomes even more crucial (Doyle, 2023; Brockman et al. 2018). With the advent of advanced digital technologies, traditional aspects of interorganizational trust are increasingly seen as insufficient for managing these complex networks (Ghobakhloo et al. 2023a, b). In this new paradigm, both the technological and human-centric dimensions of trust are vital for supporting collaboration between partners. However, existing literature has often lacked a comprehensive understanding of this trust dynamic. The study of Mubarak and Petraite (2020) presented the concept of digital trust to enhance technological learning interactions within collaborative innovation networks. Digital trust represents the integration of technology and human factors to facilitate effective learning interactions. It leverages Industry 4.0 technologies, such as big data, the Internet of Things (IoT), blockchain, and cyber-physical systems. These technologies help create an environment where trust can be established and maintained even in highly digitalized settings. Furthermore, digital trust incorporates conventional measures of trust and inter-firm trust, which complement the technology-human interactions aimed at co-innovation, knowledge sharing, and ultimately technology upgrading (Mubarak & Petraite 2020).

In collaborative innovation networks, digital trust is particularly relevant as these networks are predominantly driven by technology (Kusi-Sarpong et al. 2022; Mubarak et al. 2023a, b). Through digital trust, network partners can effectively exchange knowledge and information, enhancing the overall functioning of the innovation networks. For instance, blockchain technology can provide a transparent and immutable record of transactions and interactions, which fosters trust among partners by ensuring accountability and reducing the risk of fraud. Similarly, IoT devices can provide real-time data and insights, which enhance transparency and trustworthiness in supply chain operations. Moreover, digital trust facilitates the integration and utilization of advanced digital technologies within innovation networks. It supports the adoption of these technologies by ensuring that partners have confidence in the security, reliability, and accuracy of the digital tools and data they use. This confidence is crucial for the successful implementation of technologies such as AI and big data analytics, which rely on the availability and integrity of vast amounts of data (Ghobakhloo et al. 2024). By fostering an environment of trust, digital trust enables firms to collaborate more effectively, share sensitive information, and innovate together.

Social capital

Social capital encompasses the outcomes of human resource interactions, both tangible and intangible, and refers to the enhancement of organizational effectiveness through human relationships and interactions (Ganguly et al. 2019). It plays a crucial role in facilitating collaboration and integration among supply chain partners and global value chain partners, thereby improving technological learning and the subsequent adoption of technology. In the context of new technology implementation, social capital provides a multifaceted contribution. For example, it fosters collaboration and open communication among internal stakeholders involved in technological innovation, creating an environment where employees can build meaningful relationships and collaborate more effectively on technology implementation tasks (Al-Omoush et al. 2020). The presence of social capital brings about shared value, which aids in building inter-firm trust and breaking down knowledge silos. This trust facilitates the smooth sharing of information and knowledge pertaining to innovation implementation (Gao et al. 2019). Moreover, shared value enables collective learning, addressing the knowledge intensity required for institutionalizing new technologies, which forms the basis for technology upgrading (Galati 2022). Social capital's role in fostering collaboration and open communication is essential for the development of a cooperative culture, which is critical for successful technological learning and innovation adoption. To date, however, the enabling role of social capital for technology adoption has not been extensively studied. Insights from early literature provide valuable understanding of this interaction (Michna & Kmieciak 2020). For instance, Agostini and Nosella (2020) demonstrated that the adoption of Industry 4.0 technologies necessitates a new work environment characterized by open collaboration among inter and intra-organizational teams. Such a collaborative environment enables individuals to acquire the necessary social and technical skills for engaging with new technologies. Another empirical study on the readiness of manufacturing firms for Industry 4.0 found that cognitive, social, and structural social capital positively determine firms' ability to effectively use AI and autonomous process monitoring technologies (Ul zia et al. 2023). These findings highlight the influence of social capital on technology upgrading through the implementation and adoption of intelligent automation (Ghobakhloo et al. 2023a, b).

Moreover, social capital facilitates the integration of diverse knowledge and expertise across the supply chain and global value chain partners. This integration is essential for enhancing technological learning and adopting advanced technologies. When firms collaborate closely with their partners, they can share best practices, leverage complementary skills, and develop innovative solutions that would be difficult to achieve in isolation. Social capital thus acts as a bridge, connecting various stakeholders and enabling the flow of knowledge and information necessary for technology upgrading. Furthermore, social capital contributes to the development of a trust-based culture within organizations and across their networks. This trust-based culture encourages open dialogue, reduces resistance to change, and enhances the willingness of employees and partners to engage in continuous learning and innovation (Doyle, 2023; Petraite et al. 2022). As a result, firms are better positioned to implement and integrate new technologies, ensuring that technological advancements are effectively adopted and utilized to their full potential.

Innovation culture

Innovation culture refers to an environment that fosters innovative thinking to drive value for firms (Horibe 2016). Within a firm, a culture of innovation promotes creativity and instills confidence in creating new products, improving existing ones, reorganizing processes and workflows, and adopting novel approaches across all functional areas (Martín-de Castro et al. 2013). This culture is not only beneficial internally but also plays a crucial role in enhancing collaboration and integration among supply chain partners and global value chain partners by improving technological learning and facilitating the adoption of new technologies. When examining collaborative innovation from a cultural standpoint, various factors play an important role, including the firm's values and tangible elements, such as incentivization mechanisms, information systems, communication platforms, project decision criteria, supplier evaluation lists, and their management (Aquilani et al. 2017). These attributes are essential for practicing collaborative and open innovation as it involves "opening up" the innovation process. By fostering an innovation culture, firms can overcome cultural barriers such as a lack of openness, resistance to risk-taking, organizational inertia, and the "Not Invented Here (NIH)" syndrome, which can otherwise hinder collaborative innovation (Formica and Curley 2018).

Building a culture of innovation requires embracing values such as curiosity, creativity, flexibility, and diversity. The openness intrinsic to open innovation necessitates additional values, including trust, responsibility, authenticity, and sustainability (Edvinsson 2018; Weis 2015). These values are essential for promoting the smooth interactions and open knowledge sharing among value chain partners. There exists a strong connection between innovation culture, learning, and technological orientation (Brockman et al. 2018; Ghobakhloo et al. 2023a, b). In this context, innovation culture enables firms to develop their technological learning through participation in collaborative innovation networks, which in turn contributes to technology upgrading. Innovation culture facilitates the integration and effective collaboration among supply chain and global value chain partners by creating an environment that encourages open communication and the sharing of ideas. When firms cultivate an innovation culture, they are better equipped to engage in collaborative innovation, where the collective efforts of multiple partners can lead to the development and adoption of new technologies. This collaborative environment supports the exchange of knowledge and best practices, enabling partners to learn from each other and co-create innovative solutions (Petraite et al. 2022). Furthermore, innovation culture promotes the continuous improvement and adaptation of technologies within the supply chain. By encouraging a mindset of constant innovation and learning, firms can better respond to technological advancements and market changes. This proactive approach ensures that firms remain competitive and can effectively integrate new technologies into their operations, thereby enhancing overall supply chain efficiency and resilience (Ghobakhloo et al. 2023a, b). In addition, an innovative culture supports the alignment of technological learning with strategic business goals. When innovation is embedded in the organizational culture, firms are more likely to prioritize investments in technology that align with their long-term objectives. This strategic alignment ensures that technological advancements are not only adopted but also leveraged to drive sustainable growth and competitive advantage (Brockman et al. 2018).

Networking capability

The body of knowledge on networking capability approaches is evolving due to emergence of smart networks and digital technologies. The concept of networking capability originated in the 1990s with the emergence of the capability-centered view of networks and network-related capabilities. Notable studies by Gemunden and Ritter (1997) and Ritter (1999) aimed to bridge the gap between network theory and the capability perspective. These studies primarily addressed the limited attention given to network-related capabilities in comparison to other organizational capabilities. Drawing on the work of Mattsson (1988), Ritter (1999) emphasized the distinction between tasks necessary for managing individual relationships (dyads) and those required to manage the network as a whole. This argument underscored the importance of managing activities that encompass the entire network (Johnsen et al. 2000). Networking capability facilitates collaboration and integration among supply chain partners and global value chain partners by enhancing the technological learning necessary for adopting new technologies. Firms' networking activities extend beyond mere capability building and must manifest as recurring behavioral patterns. The network can, therefore, be viewed as an interconnected system comprising actors, resources, and activities, serving as a platform for the joint development of inter-firm resources and capabilities (Arasti et al. 2021). Networking capability plays a crucial role in cultivating relationships with network partners (Mubarik et al. 2019). Through the exchange of resources and expertise, network partners not only enhance their explicit understanding but also their tacit knowledge. This exchange fosters an environment where firms can develop an ambidextrous learning approach, which is essential for boosting their technology absorptive capacity (TAC). TAC is critical for technological learning within collaborative innovation networks, leading to the adoption and implementation of cutting-edge technologies. This, in turn, facilitates technology upgrading within these networks (Kusi-Sarpong et al. 2022).

Networking capability also enhances the ability of firms to manage and integrate activities across the supply chain and global value chains, promoting smarter and more agile networks. By effectively managing these networks, firms can better coordinate their efforts, share critical information in real-time, and respond more swiftly to market changes (Ghobakhloo et al., 2024; Ghobakhloo et al. 2023a, b). This agility is particularly important in today's fast-paced business environment, where the ability to quickly adapt to new technologies and market demands can provide a significant competitive advantage. Moreover, networking capability facilitates interactions among value chain partners by providing a structured framework for communication and collaboration. It ensures that all partners are aligned with the common goals and objectives, thereby fostering a cohesive and integrated approach to innovation and technological advancement. This alignment is crucial for the successful implementation of new technologies and for maximizing the benefits derived from these innovations. Furthermore, the development of networking capabilities helps firms to leverage their external relationships to gain access to new knowledge and technologies. This external focus complements internal technological learning efforts and enhances the overall innovation capacity of the firm. By engaging with a diverse set of partners, firms can gain insights into different technological advancements and best practices, which can be incorporated into their own operations to drive innovation and efficiency.

4.4 Conceptual Framework

This study examines the intricate dynamics of technological learning and technology upgrading within collaborative innovation networks. Central to this investigation is the role of firms' innovation capabilities in fostering technological learning. Specifically, the research delves into the relational view of innovation capabilities, which encompasses four key dimensions: social capital, trust, networking capability, and innovation culture. These capabilities are pivotal in enhancing technological learning by effectively managing relationships within the network, fostering trust to facilitate learning interactions, promoting integration within innovation networks, and encouraging collaborative innovation through an open and inclusive culture. Social capital, for instance, enhances organizational effectiveness through robust human relationships and interactions, which in turn support the smooth sharing of information and knowledge critical for technological innovation. Trust plays a vital role in ensuring seamless collaborative learning and knowledge sharing among network partners, thus accelerating technological adoption and upgrading. Networking capability further supports these processes by enabling firms to cultivate and maintain productive relationships with network partners, thereby enhancing their technology absorptive capacity (TAC). This capability ensures that firms can effectively assimilate and utilize new technologies, promoting agility and responsiveness within the network. Additionally, an innovation culture fosters an environment that encourages creativity, openness, and continuous improvement, which are essential for sustaining technological learning and innovation. The synergistic effect of these relational capabilities—social capital, trust, networking capability, and innovation culture—facilitates the successful integration and collaborative efforts within innovation networks. This, in turn, leads to effective technological learning and subsequent technology upgrading as shown in Fig. 4.1. By harnessing these capabilities, firms can navigate the complexities of collaborative innovation networks, achieve significant technological advancements, and maintain a competitive edge in the global market. Thus, the study underscores the critical importance of a relational approach to innovation

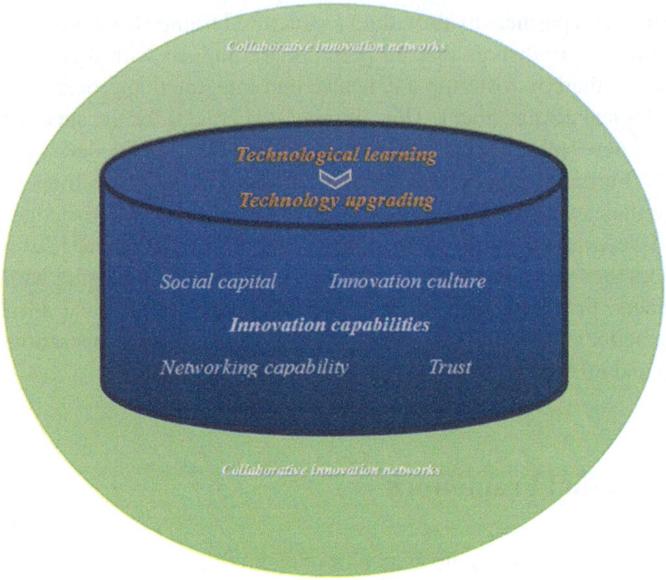

Fig. 4.1 Technology upgrading in collaborative innovation networks (Source: Authors of study)

capabilities in driving technological progress and innovation within collaborative networks. The conceptual framework of study is shown in Fig. 4.1.

4.5 Methodology

This study employs the analytical hierarchical process (AHP) to prioritize innovation capabilities through pairwise comparison, a method known for its systematic and structured approach to decision making. To execute this process, we engaged 12 experts from research, academia, and industry, following the approach used by Ghobakhloo et al. (2022). Initially, these experts were briefed on the study's objectives and provided with a detailed explanation of the innovation capabilities under consideration. To ensure clarity and mitigate any potential biases in understanding, the relational view of innovation capabilities, along with their operational definitions, were presented in accordance with the study by Mujahid et al. (2019). The innovation capabilities were then ranked based on their scores derived from pairwise comparison. This prioritization was achieved using a 3-point Likert scale questionnaire, which categorizes factors based on their perceived importance: 1 denoting "not important," 2 "somewhat important," and 3 "very important." This scale, originally developed by Saaty (1995) and applied in the AHP technique by Mubarik et al. (2018), allowed us to compute the average scores of the evaluated capabilities, thereby assessing their relative importance.

To further explore the relationships and interdependencies among innovation capabilities in relation to technology upgrading, both directly and indirectly through technological learning, we employed the interpretive structural modeling (ISM) technique. This approach helps in structuring complex relationships among variables into a comprehensible model. Two focus groups comprising the same experts were conducted to evaluate the role of innovation capabilities in technological learning and firms' technology upgrading ability.

In the first focus group, detailed instructions were provided on completing the ISM-based pairwise comparison, along with definitions of the terms used in the study. The experts carried out the pairwise comparison based on four criteria for pairwise comparison as recommended by Ghobakhloo et al. (2022; 2024), which are illustrated in Table 4.1. The criteria included the extent to which one capability influences another, the degree of dependency, the frequency of interaction, and the overall significance of each capability in the context of technological learning and upgrading. In the second focus group session, the compiled relationships, interrelationships, and interactions between capabilities were presented to the experts. This session was designed to validate and refine the initial assessments. The experts reviewed the proposed model, discussed any discrepancies, and made adjustments based on their final evaluations. This iterative process ensured the robustness and accuracy of the ISM model in representing the complex dynamics of innovation capabilities within collaborative innovation networks.

To perform the pairwise comparison for ISM, it is assumed that the hypothetical relationship matrix in Table 4.2 explains the relationships between factors F1, F2, and F3. The symbol V for the F1-F2 entry in Table 4.2 indicates that according to our hypothetical expert, factor F1 causes (enables) factor F2 as recommended by Ghobakhloo et al. (2022). However, the F1-F3 entry in this table is symbolized by A, meaning the hypothetical expert believes that factor F1 is caused (enabled) by factor F3. Finally, the F2-F3 entry in this table is symbolized by X, indicating that Factor F2 and F3 cause (enable) each other.

The binary values assigned to these four criteria (V, A, X, O) in line with Ghobakhloo et al. (2022) are further applied to compute driving and dependence powers. Based on these powers, the placement levels and interactions are mapped.

Table 4.1 Description of each sign (Source: Authors of study)

Notation(s)	Description(s)
V	The *Innovation capability* in row (i) results in (enables or causes) the *Innovation capability* in column (j)
A	The *Innovation capability* in row (i) results from (is enabled by) the *Innovation capability* in column (j)
X	The *Innovation capability* in row (i) and in column (j) have a mutual (enabling) relationship
O	The *Innovation capability* in row (i) and column (j) have no relationship

Table 4.2 Hypothetical relationship matrix of innovation capabilities (Source: Ghobakhloo et al. 2022).

Functions i/j		*j*		
		F1	F2	F3
i	F1		V	A
	F2			X
	F3			

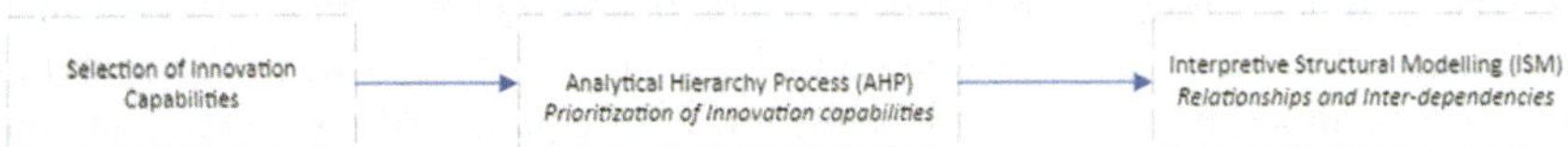

Fig. 4.2 Sequence of methodological applications (Source: Authors of study)

The binary values are assigned based on the following criteria established for ISM (Ghobakhloo et al. 2022):

- o When the entry *(i, j)* in structural self-interaction matrix is symbolized by V, the entries *(i, j)* and *(j, i)* in initial reachability matrix are, respectively, set to 1 and 0;
- o When the entry *(i, j)* in structural self-interaction matrix is symbolized by A, the entries *(i, j)* and *(j, i)* in initial reachability matrix are, respectively, set to 0 and 1;
- o When the entry *(i, j)* in structural self-interaction matrix is symbolized by X, the entries *(i, j)* and *(j, i)* in initial reachability matrix are both set to 1;
- o When the entry *(i, j)* in structural self-interaction matrix is symbolized by O, the entries *(i, j)* and *(j, i)* in initial reachability matrix are both set to 0.

The flow of methodology adopted in this study presented in Fig. 4.2 shows the prioritization and selection innovation capabilities which are then undergone to ISM to ascertain relationships and interactions between all capabilities. By integrating these methodologies, the study provides a nuanced understanding of how social capital, trust, networking capability, and innovation culture collectively enhance technological learning and facilitate technology upgrading. This methodological approach not only identifies key innovation capabilities but also elucidates the intricate web of interactions that drive technological advancement within collaborative innovation networks.

4.6 Findings

In the first step, the analytical hierarchical process (AHP) was applied to prioritize the innovation capabilities. As a result, it is revealed that innovation culture (IC) with the highest average score (IC = 2.90) holds significant potential for enhancing technological learning and technology upgrading. This highlights the pivotal role of fostering an environment that encourages creativity, openness, and continuous improvement in driving technological advancements. Then, social capital (SC) with the second highest score (SC = 2.81) was identified as a crucial capability that improves relationships within collaborative innovation networks. It facilitates both explicit and tacit knowledge transfer, thereby significantly contributing to technological learning. Further, digital trust (DT) emerged as an important factor with third ranking (DT = 2.65) in facilitating learning interactions among network partners, underscoring its role in ensuring seamless and effective communication and collaboration. Conversely, networking capability (NC) received the lowest scores (NC = 2.19), suggesting its relatively less contribution to technology upgrading.

In the second step, the interpretive structural modeling (ISM) technique was applied in this study to identify the interactions and interdependencies of innovation capabilities on technology upgrading. The experts participated in a panel meeting where they shared their perspectives on the influence of these capabilities on technological learning and technology upgrading. While opinions varied, most experts agreed that the innovation culture of firms positively influences both technological learning and technology upgrading. This consensus underscores the importance of cultivating a robust innovation culture to drive technological progress. Social capital was recognized as a capability that significantly impacts technological learning, which in turn contributes to technology upgrading. This finding highlights the role of strong, trust-based relationships in facilitating the exchange of knowledge and enhancing technological capabilities. Digital trust was identified as a capability that improves technological learning by streamlining learning interactions, thereby fostering a more collaborative and efficient innovation environment. Interestingly, while networking capability was found to have a significant impact on technology upgrading, it exhibited a relatively passive effect on technological learning. This suggests that while networking is crucial for integrating new technologies, its direct influence on the learning process may be less pronounced.

Additionally, the experts expressed their views on the direction of relationships among innovation capabilities within innovation networks. The responses were analyzed using ISM, similar to the study by Ghobakhloo et al. (2022). The analysis revealed several key interactions: the innovation culture of firms influences social capital (IC $\rightarrow$ SC), and digital trust enhances networking capability (DT $\rightarrow$ NC) within collaborative innovation networks. Moreover, innovation culture affects networking capability (IC $\rightarrow$ NC), and there is a mutual dependency between digital trust and social capital (DT $\leftrightarrow$ SC). However, digital trust and networking capability (DT x NC) were not found to be directly related to each other; rather, they act independently while contributing to improving technological learning and technology upgrading. These findings suggest that innovation capabilities should be developed and nurtured in a way that they complement each other to achieve technology upgrading. A holistic approach that integrates innovation culture, social capital, digital trust, and networking capability can create a synergistic effect, enhancing both technological learning and upgrading. The interactions between these relational categories of innovation capabilities and their role in improving technology

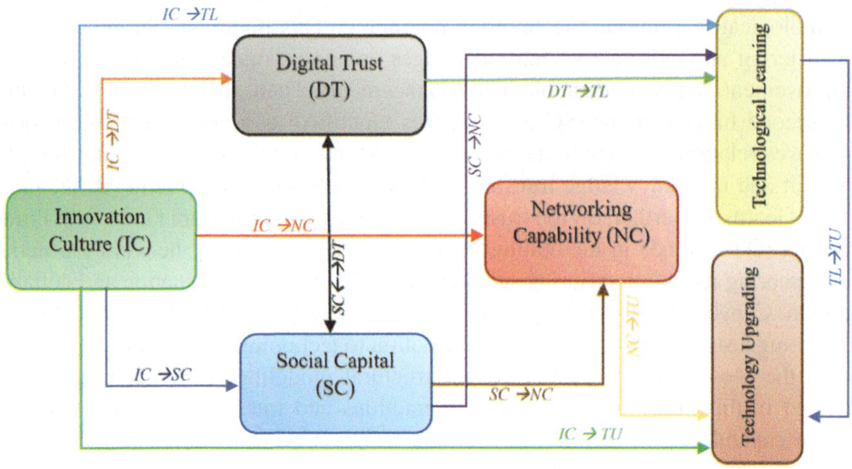

Fig. 4.3 Interactions of innovation capabilities for technology upgrading (Source: Authors of study)

upgrading directly and indirectly through technological learning are illustrated in Fig. 4.3.

4.7 Discussion and Implications

Firms focus on technology upgrading to remain competitive in the rapidly changing technological environment. Technology upgrading is achieved through constant technological learning, which involves acquiring knowledge and developing an understanding of new technologies how they can be applied to current operations. This enables firms to mitigate the effects of technological shifts and adapt to the shortened lifecycles of innovation. To facilitate this process, firms join collaborative innovation networks, such as open innovation communities and global value chain networks. Within these networks, partners exchange knowledge, resources, and expertise to drive innovation. Through active participation, firms not only gain access to valuable resources, but also enhance their technological learning through development of their innovation capabilities. This study adopted a relational view of innovation capabilities, which is most suitable for members of collaborative innovation networks. In such networks, firms interact and learn from each other through a network of relationships. The capabilities examined in this study focus on improving the innovation capabilities of firms, such as innovation culture and networking capabilities. These capabilities play a crucial role in directly and indirectly enhancing technology upgrading, as evidenced by the studies conducted by Ghobakhloo et al. (2023a, b) and Mubarak et al. (2023a, b). Moreover, interactive capabilities that are explicitly related to external partners, such as digital trust and social capital (Mubarak and

Petraite 2020; Petraite et al. 2022), are closely intertwined. Both sets of capabilities involve dependencies and relationships with members of collaborative innovation networks. Social capital was not only found to facilitate the tacit and explicit aspects of technological learning, but it also contributes to improving internal capabilities, such as the networking capability of firms through the establishment of trust and streamlined social interactions within the networks.

It is worth noting that the internally focused capabilities of firms complement their externally focused capabilities within innovation networks. Trust and social capital, encapsulated as social trust, have been shown to play a significant role in improving the effectiveness of open innovation communities, facilitating technological knowledge sharing, and promoting the adoption of advanced technology (Brockman et al. 2018; Kusi-Sarpong et al. 2022). This study also extends the notion of dynamic capabilities (Teece et al. 1997) by emphasizing the role of absorptive capability, specifically technology absorptive capacity. This capacity enables firms to effectively gain, assimilate, integrate, and utilize technological learning acquired from networks within their national innovation systems (Petraite et al. 2022). While the internally and externally focused capabilities discussed in this study may not be extensively explored in existing literature, they are considered here to clarify their role and highlight their importance in complementing other capabilities or being utilized directly within networks for managing partners.

This study recommends that firms should develop the relational and social category of innovation capabilities to strategically guide their technology upgrading endeavors. By actively participating in and integrating with collaborative innovation networks, firms can simultaneously enhance their technological learning while pursuing the strategic direction of technology upgrading. Empirical evidence by Petraite et al. (2022) supports the notion that firms, particularly those from small, open, and transitioning economies, can enhance their technology upgrading by proactively integrating with global innovation system-based international networks and consistently developing technological learning. Therefore, Research and Development intensive and technology-based firms from larger economies (e.g., UK, USA, Canada) are encouraged to associate with and derive benefits from collaborative innovation networks. Similarly, firms from smaller economies (e.g., Lithuania, Estonia, and Singapore) are advised to adopt this approach to successfully navigate the process of technology upgrading.

4.8 Limitations and Future Research

This research has specifically examined the interactive aspects of innovation capabilities that are directly linked to technology upgrading within collaborative innovation networks. It is recommended for future researchers to explore additional capabilities associated with technological learning and technology upgrading. Moreover, the study has utilized interpretive structural modeling (ISM) to establish the interactions between the interactive view of innovation capabilities and technology upgrading.

However, it is important to acknowledge that this method relies on expert knowledge and may be susceptible to knowledge and perspective biases, despite the implementation of techniques to mitigate these concerns. Therefore, it is advisable to test and validate the identified relationships using statistical modeling techniques on a large dataset of responses. Furthermore, previous research has indicated that smaller firms benefit from their participation in international networks by technological learning and upgrading from larger firms (Petraite et al. 2022). However, there are still several significant questions that remain unanswered. For example, it is unclear whether large firms actively engage in such networks and, if they do, how they can acquire technological learning from SMEs that may not possess the same level of technological advancement. Additionally, the advantages that large firms can gain in collaborative international networks in terms of technological learning and technology upgrading need to be further explored. Furthermore, existing studies have predominantly focused on technological learning and technology upgrading within technology-based firms operating in the manufacturing sectors. It is essential to investigate the role of technology upgrading in service sector firms. Many service sector firms are involved in the development and provision of innovative services, including software as a service (SaaS), among others. Hence, it is crucial to determine the relevance of technological learning for these service industries, whether it is less, more, or equally significant. Additionally, it is necessary to explore how service sector firms can contribute to technology upgrading within collaborative innovation networks, considering that technology upgrading may not be their primary focus. Although succinctly, this research also sheds light on the emerging phenomenon in quantitative research on the *"same model but different results."* In doing so, the prioritization and ranking of experts varied in AHP and ISM applications. Therefore, we suggest future researchers to further investigate this phenomenon for clarifying the differences and applicability of both techniques.

Acknowledgements The author(s) wish to declare that the manuscript has been proofread using Grammarly and ChatGPT on various paragraphs. After utilizing these tools, the authors thoroughly reviewed and modified the manuscript as necessary, taking full responsibility for its content. The work is funded by the Lithuanian Research Council (LMT) under project grant number P-PD-23–118. The study is also based on work funded by the Natural Sciences and Engineering Research Council (NSERC) of Canada under Grant No. RGPIN-2022-05008.

References

Agostini, L., & Nosella, A. (2020). The adoption of Industry 4.0 technologies in SMEs: results of an international study. *Management Decision*, 58(4), 625-643.

Al-Omoush, K. S., Simón-Moya, V., & Sendra-García, J. (2020). The impact of social capital and collaborative knowledge creation on e-business proactiveness and organizational agility in responding to the COVID-19 crisis. *Journal of Innovation & Knowledge*, 5(4), 279-288.

Amis, J. M. (2018). Understanding organization change and innovation: A conversation with Mike Tushman. *Journal of Change Management*, 18(1), 23-34.

Aquilani, B., Abbate, T., & Codini, A. (2017). Overcoming cultural barriers in open innovation processes through intermediaries: A theoretical framework. *Knowledge Management Research & Practice*, 15(3), 447-459.

Brockman, P., Khurana, I. K., & Zhong, R. I. (2018). Societal trust and open innovation. *Research Policy*, 47(10), 2048-2065.

Dodgson, M. (1991). Technology learning, technology strategy and competitive pressures. *British Journal of Management*, 2(3), 133-149.

Dodgson, M., Ash, S., Andrews, J., & Phillips, N. (2022). Managing technology-enabled innovation in a professional services firm: a cooperative case study. *Academy of Management Discoveries*, 8(4), 509-530.

Dodgson, M. (2018). *Technological collaboration in industry: strategy, policy and internationalization in innovation* (Vol. 11). Routledge.

Doyle, J. (2023). Social trust, cultural trust, and the will to sacrifice for environmental protections. Social Science Research, 109, 102779.

Edvinsson, L. (2018). The Culture of Open Innovation–A New Twilight Zone? In Exploring the Culture of Open Innovation (pp. 55–67). Emerald Publishing Limited.

El-Awad, Z., Gabrielsson, J., & Politis, D. (2017). Entrepreneurial learning and innovation: The critical role of team-level learning for the evolution of innovation capabilities in technology-based ventures. *International Journal of Entrepreneurial Behavior & Research*, 23(3), 381-405.

Formica, P., & Curley, M. (2018). In search of the origin of an 'open innovation' culture. In Exploring the culture of open innovation (pp. 1–54). Emerald Publishing Limited.

Galati, F. (2022). Blockchain adoption in supply networks: a social capital perspective. *Supply Chain Management: An International Journal*, 27(7), 17-32.

Ganguly, A., Talukdar, A., & Chatterjee, D. (2019). Evaluating the role of social capital, tacit knowledge sharing, knowledge quality and reciprocity in determining innovation capability of an organization. *Journal of Knowledge Management*, 23(6), 1105-1135.

Gao, Y., Liu, B., Yu, L., Yang, H., & Yin, S. (2019). Social capital, land tenure and the adoption of green control techniques by family farms: Evidence from Shandong and Henan Provinces of China. *Land Use Policy*, 89, 104250.

Ghobakhloo, M., Iranmanesh, M., Mubarak, M. F., Mubarik, M., Rejeb, A., & Nilashi, M. (2022). Identifying industry 5.0 contributions to sustainable development: A strategy roadmap for delivering sustainability values. *Sustainable Production and Consumption*, 33, 716-737.

Ghobakhloo, M., Iranmanesh, M., Mubarik, M. S., Mubarak, M. F., Amran, A., & Khanfar, A. A. (2024). Blockchain technology as an enabler for sustainable business ecosystems: A comprehensive roadmap for socioenvironmental and economic sustainability. Business Strategy & Development, 7(1), e319.

Ghobakhloo, M., Asadi, S., Iranmanesh, M., Foroughi, B., Mubarak, M. F., & Yadegaridehkordi, E. (2023). Intelligent automation implementation and corporate sustainability performance: The enabling role of corporate social responsibility strategy. *Technology in Society*, 74, 102301.

Ghobakhloo, M., Iranmanesh, M., Tseng, M. L., Grybauskas, A., Stefanini, A., & Amran, A. (2023). Behind the definition of Industry 5.0: a systematic review of technologies, principles, components, and values. *Journal of Industrial and Production Engineering*, ahead-of-print.

Horibe, F. (2016). Creating the innovation culture. eBook Partnership.

Johnsen, T., Wynstra, F., Zheng, J., Harland, C. & Lamming, R. (2000), "Networking activities in supply networks", *Journal of Strategic Marketing*, 8(2), 161-181.

Kruss, G., McGrath, S., Petersen, I. H., & Gastrow, M. (2015). Higher education and economic development: The importance of building technological capabilities. *International Journal of Educational Development*, 43, 22-31.

Kusi-Sarpong, S., Mubarik, M. S., Khan, S. A., Brown, S., & Mubarak, M. F. (2022). Intellectual capital, blockchain-driven supply chain and sustainable production: Role of supply chain mapping. *Technological Forecasting and Social Change*, 175, 121331.

Martín-de Castro, G., Delgado-Verde, M., Navas-López, J. E., & Cruz-González, J. (2013). The moderating role of innovation culture in the relationship between knowledge assets and product innovation. *Technological Forecasting and Social Change*, 80(2), 351-363.
Mattsson, L.-G. (1988), "Management of strategic change", in a "Markets-as-Networks" Persective, in the Management of Strategic Change, in Pettigrew, A.M. (Ed.), Basil Blackwell, Oxford, 234–256.
Michna, A., & Kmieciak, R. (2020). Open-mindedness culture, knowledge-sharing, financial performance, and industry 4.0 in SMEs. *Sustainability*, 12(21), 9041.
Mohrman, S. A., & Von Glinow, M. A. (1990). High technology organizations: Context, organization and people. *Journal of Engineering and Technology Management*, *6*(3-4), 261-280.
Mubarak, M. F., & Petraite, M. (2020). Industry 4.0 technologies, digital trust and technological orientation: What matters in open innovation? *Technological Forecasting and Social Change*, 161, 120332.
Mubarak, M. F., Petraite, M., Rasli, A., & Shabbir, M. (2023). Capability framework to support supply chain open innovation networks. In Blockchain Driven Supply Chain Management: A Multi-dimensional Perspective (pp. 119–134). Singapore: Springer Nature Singapore.
Mubarak, M. F., Petraite, M., Rasli, A., & Shabbir, M. (2023). Capability Framework to Support Supply Chain Open Innovation Networks. In *Blockchain Driven Supply Chain Management: A Multi-dimensional Perspective* (pp. 119–134). Singapore: Springer Nature Singapore.
Mubarik, M. S., Chandran, V. G. R., & Devadason, E. S. (2018). Measuring human capital in small and medium manufacturing enterprises: what matters?. Social indicators research, 137, 605-623.
Mubarik, M. S., Naghavi, N., & Mubarak, M. F. (2019). Impact of supplier relational capital on supply chain performance in Pakistani textile industry. *Asian Economic and Financial Review*, 9(3), 318.
Mujahid, S., Mubarik, S., & Naghavi, N. (2019). Prioritizing dimensions of entrepreneurial ecosystem: a proposed framework. Journal of Global Entrepreneurship Research, 9, 1-21.
Oqubay, A., & Ohno, K. (2019). *How nations learn: Technological learning, industrial policy, and catch-up* (p. 368). Oxford University Press.
Petraite, M., Mubarak, M. F., Rimantas, R., & Von Zedtwitz, M. (2022). The role of international networks in upgrading national innovation systems. *Technological Forecasting and Social Change*, 184, 121873.
Qin, J., van der Rhee, B., Venkataraman, V., & Ahmadi, T. (2021). The impact of IT infrastructure capability on NPD performance: The roles of market knowledge and innovation process formality. *Journal of Business Research*, 133, 252-264.
Radosevic, S., & Yoruk, E. (2018). Technology upgrading of middle income economies: A new approach and results. *Technological Forecasting and Social Change*, 129, 56-75.
Ritter, T. (1999), "The networking company: antecedents for coping with relationships and networks effectively", *Industrial Marketing Management*, 28(5), 467-479.
Rubera, G., Chandrasekaran, D., & Ordanini, A. (2016). Open innovation, product portfolio innovativeness and firm performance: the dual role of new product development capabilities. *Journal of the Academy of Marketing Science*, 44, 166-184.
Saaty, T. L. (1995). Transport planning with multiple criteria: The analytic hierarchy process applications and progress review. Journal of advanced transportation, 29(1), 81-126.
Shabbir, M., Petraite, M., & Mubarak, M. F. (2024). Trajectory of Technology Upgrading: Intellectual Structure and Future Research Directions. 2024 IEEE Technology and Engineering Management Society (TEMSCON LATAM), 1-6.
Teece, D. J., Pisano, G., & Shuen, A. (1997). Dynamic capabilities and strategic management. *Strategic Management Journal*, 18(7), 509-533.
ul zia, N., Burita, L., & Yang, Y. (2023). Inter-organizational social capital of firms in developing economies and industry 4.0 readiness: the role of innovative capability and absorptive capacity. *Review of Managerial Science*, 17(2), 661-682.
Weis, B. X., & Weis, B. X. (2015). Innovation Culture and Innovation Management. From Idea to Innovation: A Handbook for Inventors, Decision Makers and Organizations, 103–165.

Zahra, S. A., Ireland, R. D., & Hitt, M. A. (2000). International expansion by new venture firms: International diversity, mode of market entry, technological learning, and performance. *Academy of Management Journal*, 43(5), 925-950.

Chapter 5
Ethical Issues and Role of Smart Supply Chain Management

Hamza M. Dawood and Nimra Afzal

5.1 Introduction

Increasing consumer demands and concerns have recently heightened awareness of unethical behavior inside supply chains. The relentless demand of customers for affordable goods and the widespread availability of genuine-looking counterfeit products have emerged as significant apprehensions. Many retailers are emphasizing the ethical issues of manufacturers and employees (Ucaryilmaz Deibel 2022). As a result of the phenomenon of globalization and the practice of offshoring, organizations are competing within supply chains or interconnected demand networks. Alongside achieving notable economic achievements, corporations are facing a growing responsibility for their suppliers' and partners' environmental and social performance. Multiple stakeholders, such as employees, executives, socially responsible organizations, local communities, governments, and non-governmental organizations, impose these pressures (Mc Loughlin et al. 2023). The scope of supply chain ethics includes discrimination, working conditions, and environmental responsibilities. As the supply chain becomes more robust, businesses have been forced to prioritize operational ethics (Wainwright and Muñoz 2020). Environmental and ethical considerations need to be addressed throughout supply chains, which oversee the movement of materials and information and encompass the regional strength of supply chains, regional activities, and the manufacturing and public sector operations of small and medium-sized businesses (Yang 2022). Sustainability or ethical issues may damage brand equity and sales, making supply chain performance a critical competitive factor (Tsai et al. 2021).

The chapter examines ethical issues in supply chains driven by emerging developing technologies (EDTs) like blockchain, AI, intelligent sensors, etc. (Mishra et al.

H. M. Dawood (✉) · N. Afzal
School of Management and Economics, University of Electronic Science and Technology of China, Chengdu, China
e-mail: 202014150117@std.uestc.edu.cn; hamzamd@foxmail.com

M. S. Mubarik and S. A. Khan (eds.), *Smart Supply Chain Management*,
https://doi.org/10.1007/978-981-96-1333-5_5

2023). EDTs improve supply chain management (SCM) but must be utilized ethically. Ethical concerns in supply chains may help solve these difficulties and promote smart, ethical supply systems. EDTs instrument, interconnect, integrate, intelligent, automate, and innovate the smart supply chain, improving its performance (Kunduzovihc 2022). The smart supply chain has great potential to improve supply chain resilience, sustainability, agility, and responsiveness. It has vast applicability, but few companies have implemented it even though its importance has increased in recent times (Rogerson and Parry 2020). The prevailing ethical concerns encompass bias and prejudice, harassment, unscrupulous bookkeeping, misuse of technology, data privacy, health and safety, favoritism and nepotism, and harm to the organization's environment where workplaces commonly experience these difficulties. Gaining a comprehensive understanding of these challenges enables firms to effectively manage them. Acquiring the ability to identify and mitigate these obstacles could assist coworkers in focusing on the advancement of the organization rather than dealing with remedial tasks (LeBaron 2021). The theory of traditional supply chains, which focuses on obtaining things from the cheapest vendors, has significant shortcomings that prevent them from fully achieving their potential. Supply chain ethics examines and establishes optimal methods employed by worldwide manufacturers, merchants, and logistics organizations through the examination of real-life examples, surveys, and original research. The study examines the phenomenon known as the "triple advantage" that companies get by integrating financial profitability, environmentally sustainable practices, and positive social impacts. Supply chain strategies that disregard two or more criteria are destined to fail (Rounaghi et al. 2021). The following are further motivations to fill the gap in ethical SCM research:

Undiscussed Ethical Issues: The section will cover ethical concerns in SCM, although it does not address typical company ethical difficulties. The chapter addresses child labor, working conditions, and environmental sustainability.

Stakeholder Perspectives: This part will delve into the various stakeholder perspectives and their respective roles, responsibilities, and expectations regarding ethical behavior in the supply chain. The ethical issues in SCM involve many stakeholders, including suppliers, manufacturers, distributors, retailers, consumers, and local communities. Understanding the diverse perspectives of these stakeholders is crucial for developing effective ethical strategies.

Ethics discussion:

The section briefly discusses how ethical concerns affect stakeholders, but it does not address how they affect a company's reputation, brand, and financial performance. Ethical SCM business case is also discussed; however, SCM rules are not covered. Possible additions include the United Nations Global Compact, the OECD Guidelines for Multinational Enterprises, and the UK Modern Slavery Act. This chapter does not address the topic of ethical leadership in supply chain management. It also presents an opportunity to discuss how a company's leadership influences its culture.

By addressing these gaps, the chapter aims to contribute to the existing knowledge on ethical issues in supply chain management. It will provide valuable insights for researchers, practitioners, and policymakers, enabling them to make informed decisions and develop strategies to foster ethical behavior in global supply chains.

The opening part of the book chapter focuses on introducing the ethical concerns linked with the supply chain. The second section of the book chapter discusses the recent literature on ethics and supply chains. The third section examines the various dimensions of smart supply chain ethics and their potential implications. Finally, this chapter provides implications and concludes in the fourth and fifth sections.

5.2 Literature Review

This section will discuss ethical issues, levels, principles, and types of ethical issues. Also, this part of the book chapter will discuss the relevant examples of ethical issues in smart supply chains.

5.2.1 Ethical Issues

An organizational systems approach to the design of ethical organizations includes both theoretical and practical knowledge that may be used to assist in the creation of organizations with high integrity that are also highly effective. The book advises how to structure companies to encourage moral conduct and reduce hazards (Day et al. 2021). Ethical challenges arise in the workplace when a moral dilemma needs resolution. For example, it is a circumstance when the validity of a commonly held moral principle is questioned. When a person faces a situation where their actions run counter to the values of their community, they face an ethical problem. Conflicts like this may include individuals and businesses whose actions may be morally questionable (Shaun 2022). For Instance, IKEA furniture, being a multinational corporation, cannot avoid adhering to the ethical standards of any nation in which they have operating units.

5.2.2 The Impact of Supply Chain Ethics on Business

Despite being frequently undervalued, supply chain ethics are crucial to a company's success. Even one instance of unethical behavior can harm a company's reputation and bottom line. Consumers want evidence that their purchases support sustainable and ethical production methods. Customers will be lost if a company's actions do not align with its advertising. The positive news is that 83% of supply chain professionals know this and consider supply chain ethics important or essential (Mubarik et al. 2021). However, this information is not always applied to developing ethical supply chains, as "creating ethical supply chains is not only a requirement; it is also an opportunity." In addition, executives need an understanding of the supply chain that serves their organization (Staff 2019). While ethical dilemmas for business leaders have always existed, their character and importance have shifted in the digital age. In the twenty-first century, there are a plethora of ethical concerns that must be addressed as the complexity of business management rises. Therefore, these reviews have been conducted as an in-depth understanding of ethical issues and creating mitigation measures is essential (Nuseir and Ghandour 2019).

5.2.3 Levels of Ethical Issues

Business managers and leaders often face ethical dilemmas on a group, national, and global scale. With the rise in popularity of the digital world, social media, and the internet, three levels of influence have emerged to depict their impact: macro (about national and societal pressure), business, and individual. The company's operations are encompassed at the systemic or macroscopic level. It could pertain to a particular city, region, or even a country. Company or corporate ethics refers to the set of principles and guidelines that govern the behavior and actions of an organization. Ethical dilemmas can arise for employees, independent contractors, suppliers, rivals, coworkers, and superiors (Tups and Dannenberg 2023). When confronted with a moral conundrum, we face an ethical difficulty. Therefore, ethical dilemmas frequently counter the values of a given culture. Common ethical dilemmas are discrimination, harassment, unethical accounting practices, technology misuse, data privacy concerns, health and safety issues, favoritism, and nepotism. These concerns are often voiced while at work (Burtler 2022). Understanding these difficulties is critical to handle them when they develop in the company. Knowing how to recognize and prevent these problems before they become problems allows colleagues to concentrate on company success and development rather than correction (ADP 2022).

5.2.4 Common Ethical Issues

Many high-profile cases of sweatshop labor conditions in suppliers to global brands like Nike and Walmart have raised awareness of human rights issues in the workplace, including child and forced labor, health and safety, discrimination, pay, and working conditions. Numerous companies' global supply chain practices have been criticized as exploitative, putting additional emphasis on management and worker rights in the workplace (Ullah et al. 2021). Combining this with ethical practices is encouraged to avoid a reputation for exploitation anywhere in the supply chain. Social and environmental regulations benefit the economy by promoting long-term viability through environmental sustainability, increasing worker health and productivity, and decreasing medical costs (Zaman et al. 2023a). The ethical issues in the supply chain are multifaceted, ranging from exploitative labor practices and environmental concerns to bribery and corruption. These challenges demand transparent and responsible actions from companies to ensure fair treatment of workers, sustainable sourcing, and accountable business conduct (Martins et al. 2023).

5.3 Principles of Business Ethics

The basic principles of business ethics are Leadership, Honesty, Loyalty, Fairness, Compassion, Responsibility, Respect for the law, Respect for others, and Integrity (Fig. 5.1), which are detailed below:

1. Leadership involves intentionally embracing, integrating, and exemplifying the remaining eleven principles in both one's professional and personal lives.
2. Integrity encompasses honesty, trustworthiness, and reliability. An individual with integrity continually upholds moral principles and actively seeks personal growth.
3. Respecting individuals is essential for establishing an ethical working environment. Every individual is entitled to receive treatment characterized by benevolence, empathy, comprehension, and all other fundamental entitlements inherent to human beings.
4. Honesty creates an ethical climate that necessitates absolute transparency in all transactions. Discovering solutions necessitates the communication and acceptance of negative information in the same way as positive information.

Fig. 5.1 Principle of business ethics. Source: Author

5. Respecting the law and enforcing all relevant regulations is a crucial aspect of ethical leadership. Leaders should prioritize adherence to the law when uncertain, rather than attempting to take advantage of a technicality.
6. Enabling employees to assume responsibility and accountability for their tasks enhances an organization's culture of ownership.
7. Businesses should exhibit compassion toward their employees, customers, vendors, and community members.
8. Fairness refers to the state in which all individuals are granted equal rights and privileges. It is unjust if your daily schedule or behavior causes discomfort to others or prioritizes your or your company's interests over others; basic politeness and consideration.
9. Loyalty is cultivated by leaders to instill trust and inspire confidence among both staff and customers. They motivate employees and executives to consistently carry out work operations with efficiency, ensuring their dedication to optimal methods (TWIN 2022).

5.3.1 Types of Ethical Issues

Every industry, area, region, country, and organization's policies and leadership present distinct ethical issues. Although not every situation will involve a critical ethical dilemma, most will involve at least some. Such situations have uncovered complications such as cybercrime, overcharging, privacy risks, and disclosures on social media, as well as legal obligations under current legislation, workplace safety, avoiding child labor, cultural discrimination, and bribery. Moreover, in management and commerce, there are many instances of fraud, misrepresentation, phony refunds, etc. (Nuseir and Ghandour 2019).

5.3.1.1 Ethical Issues in the Workplace

Workplace ethics refers to rules corporations and other organizations may abide by, including ethical and legal guidelines. Workplace ethics have a significant impact on both the way a company handles its customers and its employees (Carroll and Brown 2022). These notions influence the relationships between a company and its customers and staffs. For instance, if a firm has alleged that it is devoted to acting ethically and honoring its promises, then that company may build robust support mechanisms that address this particular policy and assist consumers in feeling as if they are being cared for (Vitolla et al. 2021). From here, the same organization may build incentive programs to motivate workers to generate high-quality work and ensure they follow through on completing such promises when targets are accomplished. These programs may also be designed to ensure that employees produce high-quality work (Krosel 2022). Ethical issues in the workplace can be tricky problems that need to be handled thoughtfully. These issues involve making decisions

Fig. 5.2 Types of ethical issues in the workplace, Source: (ADP 2022), Readjusted by Author

about what is right and wrong in the context of work and business. Some common ethical issues in the workplace include misuse of leadership positions, unethical accounting, health and safety, technology, discrimination, religion, and elderly staff (Fig. 5.2).

1. *Misuse of Leadership Position Frequently:* Harassment and discrimination are forms of power abuse. However, leaders also have the power to abuse their positions by demanding inappropriate favors, setting impossible standards for performance, or punishing employees who fall short.
2. *Unethical Accounting:* Unfair accounting practices may be used by publicly listed corporations to make their financial situation seem better than it is.
3. *Health and Safety:* Legal safeguards are often granted to issues pertaining to health and safety. Certain enterprises may engage in unethical activities to reduce expenses and enhance production. Disregarding worker safety may result in physical injuries and psychological hazards, such as job insecurity or diminished autonomy, which may induce workplace stress.
4. *Technology, Privacy, and Social Media:* Improper use of social media can have severe consequences for online and physical privacy and security. In recent years, workers have seen a blurring of lines between their professional and personal lives. This is primarily due to technological advancements. To combat discrimination, HR professionals and senior managers should educate their employees on these issues and promote a positive workplace culture.
5. *Discrimination and Favoritism:* Discrimination is defined as any action that results in unfavorable treatment of an employee. When a business gives a job to a relative or friend of an executive, it is called nepotism. When a manager prefers

one worker over another for subjective reasons, this is known as favoritism. Both nepotism and favoritism are morale-killers in the workplace. Promotions and other remunerations frequently necessitate a substantial increase in effort.
6. *Religion and Elderly Staff:* Employees' religious beliefs should not influence how any organization member behaves toward them. Organizational and internal policies may result in discrimination against elderly personnel (ADP 2022).

5.3.2 Examples of Ethical Issues

One typical example of an ethical problem is when a company employs detrimental or unscrupulous practices to manufacture its products. This may include child labor, pollution, and hazardous working conditions. Customers and workers may be oblivious to these practices, and they may cause them injury. If a business is discovered to be engaging in unethical behavior, it may encounter legal penalties or negative publicity (Taiyabmustafiz 2022). In a practical business example, Enron, a large multinational energy corporation, submitted inaccurate financial statements for years. Enron's auditor had conspired with the accounting firm Arthur Andersen LLP to approve fraudulent financial statements. Both companies went down when the truth came out. Thousands of jobs were lost, and Enron stockholders suffered major financial losses (Holton 2020).

According to the U.S. Department of Labor, a recent case where more than 300 juveniles, including two 10-year olds, were working in violation of federal labor laws at McDonald's franchise restaurants in Kentucky and other states has garnered a potential ethical dilemma. Three separate licensees operating 62 McDonald's locations in Indiana, Kentucky, Maryland, and Ohio violated federal labor laws by employing 305 minors to work more than the legal maximum hours and to perform unlawful duties. McDonald's USA has made its policies regarding minors visiting a parent or guardian at work clear to its employees. The Labor Department determined that 242 minors aged 14 and 15 could work more than the permissible hours, totaling $143,566 in fines. Child labor laws protect children's health, safety, and education, yet such cases are exposed regularly (CNBC 2023). As more ethical issues are explored in the workplace, it is essential to recognize that similar concerns also extend to SCM. Ensuring responsible and fair practices throughout the supply chain becomes crucial to address labor conditions, environmental impact, and fair trade, fostering transparency and ethical decision-making from the initial source to the final consumer. By addressing these concerns collectively, businesses can strive to create a positive impact within their organizations and entire supply chain network (Kouhizadeh et al. 2021).

5.3.3 Supply Chain Management (SCM): Natures and Ethics

The concept that "the whole can be greater than the sum of its parts" is at the heart of SCM, which places a premium not only on the formation of cooperative relationships and the inculcation of mutual trust among the many organizations involved but also on the acknowledgment of the synergistic impact (Waters 2021). Not only does living an honest and moral life count toward one's ethical duties, but so does making sensible choices when confronted with unforeseen challenges. According to Martin et al. (2021), it is our ethical responsibility to support our coworkers in the challenging job that they are doing and to help them address technological issues in an educated manner.

5.3.3.1 Organizational Culture as a Moderator in Business Ethics

Organizational culture refers to the complex and interconnected system of behavior, attitude, and fundamental beliefs that are collectively shared by members of an organization. Individuals residing within a society or organization contribute to the unique characteristics of the culture and maintain its principles that characterize it. This demonstrates that values are strong convictions held throughout time and pertain to the company's objectives (Kwarteng et al. 2022). The company's objectives significantly impact the values it maintains, determining whether an activity is executed favorably or adversely; the foundational elements of a culture are essential for generating value inside any business. Previous researchers highlight the features of shared perceptions and shared ideas and relate this to how people conceive the world around them. It has been emphasized that it is important for businesses to comprehend the connection between ethical ideals and the corporation's culture (Naveed et al. 2022). Within the realm of SCM, a more effective alignment of purchaser expectations and supplier operations may be achieved by managing supplier relations and suppliers using metrics, programs, and strategies. This management's overarching objective is to cut down on the total risk to the buyer posed by a breach in corporate integrity anywhere along the supply chain (Jiang et al. 2023).

5.3.4 Smart Supply Chain Management (Smart SCM)

Smart supply chains also referred to as intelligent or digital supply chains employ advanced technologies like the Internet of Things (IoT), artificial intelligence (AI), blockchain, big data analytics, and cloud computing to efficiently and effectively manage the movement of goods, information, and funds across the supply chain (Charles et al. 2023). These technologies bring numerous benefits and efficiencies to the supply chain process but also raise several ethical concerns that must be addressed. Among the most important ethical concerns in automated supply chains are the following:

- *Data Privacy and Security:* Implementing intelligent supply chains requires collecting and analyzing enormous quantities of data from numerous sources.
- *Transparency and Accountability:* Multiple stakeholders and interconnected systems are frequently involved in intelligent supply chains. Transparency must be maintained throughout the supply chain so all participants can be held accountable for their actions.
- *Worker Rights and Labor Conditions:* As supply chains become increasingly automated and digitized, there is a growing concern about the possibility of laborers losing their jobs.
- *Environmental Impacts:* While intelligent supply chains can help optimize transportation and reduce waste, environmental effects may be associated with the technology's production and disposal. Ethical considerations must include an evaluation of these technologies' global environmental impact.
- *Fair Trade and Fair Labor Practices:* Intelligent supply chains can involve intricate global supplier networks. To prevent exploitation and unethical labor practices, it is essential to ensure equitable trade practices, fair remuneration, and secure working conditions for all workers in the supply chain.
- *Bias and Discrimination:* Artificial intelligence and machine learning algorithms are crucial for demand forecasting, inventory management, and decision-making in inefficient supply chains.
- *Supply Chain Control and Dependence:* As supply chains become more interconnected and dependent on digital technologies, there is a risk of centralization and control by a few influential parties. This can create dependency issues and impede equitable market competition.
- *Data Ownership and Intellectual Property:* The data generated and disseminated across intelligent supply chains may be valuable and exclusive. Data ownership, sharing agreements, and the potential for intellectual property disputes raise ethical concerns (Zhang et al. 2023).

To establish a smart supply chain, it is necessary to adopt intelligent technologies and ensure that the organization's procedures and culture are in line with data-driven decision-making and collaborative practices throughout the supply chain network. Although initial investments may be necessary, the long-term benefits can significantly enhance a company's competitive advantage in the global economy. These ethical challenges must be addressed by a collective effort involving enterprises, governments, regulatory agencies, and other stakeholders. Creating and following comprehensive ethical rules, legislation, and industry standards can help establish smart supply chains in a socially responsible and beneficial way (Di Vaio et al. 2023). Smart supply chain management with an eye on environmental sustainability involves minimizing negative effects on the natural world from the production and consumption of commodities. The product recovery function (recycling, remanufacturing, or re-use) or the product design function alone (e.g., design for the environment) have typically received the most attention in this sector (Hazen et al. 2021). Today, nearly every organization has implemented a corporate ethics policy, thanks in part to advancements in technology and digital communication. These advancements have made it simpler to identify and bring to light instances of ethical misconduct. Businesses are allocating additional resources toward corporate ethics to prevent unfavorable consequences. A survey of accountants found that 55% of respondents anticipate a rise in the significance of corporate ethics within the next three years (Zaman et al. 2023a). Businesses adopt established procedures and hire the necessary people to create ethical workplaces. "High integrity and honesty" are the second most crucial traits for business executives, according to a recent survey. Contemporary administrators need to understand commercial success and corporate ethics (Schroeder 2021). Unfortunately, forced labor has long been a part of supply chains as companies seek low-wage labor markets in countries with less formal or even formal human rights protections (Ferrell and Fraedrich 2021).

5.3.4.1 Ethical Responsibility and Its Impact on Smart Supply Chain

Stakeholders demand that enterprises acknowledge, oversee, and restrict their social and environmental impacts on the globe. As a result, the demands for ethical practices in supply chain management provide a substantial challenge for businesses across various industries and sizes. The establishment and maintenance of ethical standards in supply chain management can help alleviate human rights and environmental concerns. Another potential result is the acquisition of enduring advantages that positively impact the financial performance of the organization (House 2022). As a result, the relationship between ethical responsibility and the efficiency of supply chains has become a major focus of academic research and practical application in the corporate world. According to Yusuf et al. (2014), Beamon provides a clear definition of supply ethics as the act of delivering goods and services to consumers while following a set of ethical principles. Aytan and Sayan (2019) define it as "managing the optimal flow of high-quality, value-for-money materials, components, or services from a suitable set of innovative suppliers in a fair, consistent, and reasonable manner that meets or exceeds societal norms, even though it is not legally required." This is a more in-depth definition of the phrase "manage the optimal flow of high quality, value-for-money materials, components, or services." This highlights the need to maintain a moral compass across the product life cycle, from development to sale (Mella 2019). Establishing supply chains that conform to ethical norms is not only a necessity but also a chance for growth and advancement. Moreover, for business executives to fully capitalize on that potential, they must maintain close communication with the supply chain linked to their corporation (House 2022).

5.3.4.2 Ethical Issues in Supply Chain

Many ethical issues must be addressed to meet supply chain participants' expectations. Ethical problems revolve around whether behaviors are right or wrong, good or terrible, beneficial or harmful, and deserving of praise or punishment (Alam et al. 2023). Achieving a high level of channel integration is a complex endeavor, and the reasons for this are apparent. First and foremost, each company has its purpose, goals, objectives, and strategies. Second, acknowledging and accepting reciprocal interdependence within the supply chain contradicts our natural self-interest-seeking instincts. The following arguments examine some concerns:

- Adherence to accounting standards and regulations is a critical regulatory concern.
- Any supply chain member engaging in corruption that may affect others is wrong.
- All supply chain stakeholders must ensure that regulatory rules regarding research and development are followed rigorously.

Aside from these complexities, many regulatory instruments are directly or indirectly related to logistics and transportation. These include legislation, directives, technical specifications, prohibitions, and rules (ADP 2020).

5.3.4.3 Emerging Ethical Issues in the International Supply Chain

Regardless of the level of technological integration and utilization in business, the effectiveness of global supply chains will always depend on the ability of individuals to build relationships. Buyers and suppliers who value ethical business practices and aim to collaborate with companies that uphold high ethical standards may encounter difficulties and inconveniences when dealing with local supply chains that have a worldwide presence (Zaman et al. 2023b). Climate change may heighten problems such as wage inequality, hazardous working environments, and gender bias in the workplace. 83% of supply chain managers consider ethics to be of high or extremely high importance. Ethical supply chain management is driven by consumer demand, financial success, and public scrutiny (Gilpin 2022). When multiple companies are engaged in a supply chain activity, the issue of fairness arises, distinguishing between those that are just and those that are not. When multiple participants in the supply chain hold divergent perspectives on the definition of "fair play," it is inevitable that tensions will emerge. Different parties may perceive some business practices as either ordinary or unfair competition (Manning et al. 2022). Here are some moral challenges that supply chain participants may encounter:

- *Producers:* A manufacturer may face several moral dilemmas along the supply chain. A corporation may deviate from ethical business standards in many areas, including employment conditions, wages, usage of child labor, production methods, waste management, and use of natural resources.
- *Suppliers:* Suppliers maintain lesser inventory management standards, or the quality of goods acquired is of inferior quality, for reasons ascribed to cost savings and increased profit margins.
- *Retailers:* Similar unethical actions have been documented with merchants regarding product packaging or labeling (Manning et al. 2022).

Fig. 5.3 How to create supply chains that are ethical and sustainable in six steps. Source: Nilsson and Göransson (2021)

The current body of literature about ethical concerns in the supply chain is assessed to demonstrate the need to investigate the impact of ethical issues on the effective management of supply chain uncertainty. Numerous authors have engaged in discussions over the significance of ethics in the realm of business (Settembre-Blundo et al. 2021). For example, Friedman claimed there is no urgency to behave ethically unless it suits the company's aims, but others emphasize the relevance of ethics only on moral reasons. It has also been stated that the importance of business ethics is determined by one's position and participation in the supply chain (Shetty and Bhat 2022). A case study by Radanliev et al. (2021) revealed several significant unethical practices such as collusion or parallel contact between businesses delivering the same product to clients; corporations influencing government legislation in their favor; and firms "abusing" their influence at the cost of other enterprises (Fig. 5.3).

5.3.5 Establishing Ethical Sustainable Supply Chains

1. *Managing supplier communities:* Managing unethical activities is a constant process; therefore, businesses should look at methods to boost cooperation within their supply chains daily. Increased adoption and adherence to ethical sourcing practices may be promoted via collaboration with efficient trade partners. Supply chain members' contact information should be kept up-to-date by companies.
2. *Gather Ethical Insights:* Many companies still have difficulty keeping tabs on their trading partners' output and getting a feel for the ethical "pulse" of their supply chains.
3. *Securing partnerships:* Once a supplier has been selected, they must communicate effectively with your business. As a result, there is more faith in and less danger associated with dealings between trade partners.

4. *Create a digital supply chain:* To create a digital supply chain, organizations must acquire the requisite trading partners and connect them to the company's operations electronically.
5. *Monitor shipment provenance:* To build trust and keep the company's goodwill, it is crucial to know where the product's components came from, so keeping tabs on the origin of shipments is important.
6. *Identify trustworthy suppliers:* Before launching an ethical supply chain strategy, businesses must find reliable suppliers who share their commitment to morality. They could search for acceptable partners based on factors like whether or not the business uses conflict-free minerals in its products or has fair labor standards. Businesses must trust their suppliers to ensure all employees are treated fairly across the supply chain (Nomi 2022).

5.3.5.1 Creating an Ethically Accountable Supply Chain

SCM has undergone significant adjustments as a result of current developments in health care and logistics. As the potential for disruption rises, the ethical standards of the supplier decline (Holgado and Niess 2023). Modern slavery encompasses various manifestations, such as uncompensated or unbenefited labor, inadequate workers' rights, and instances of time-sheet fraud resulting in incorrect remuneration (Achillles 2022). Most ethical firms strive to ensure that optimal procedures are adhered to throughout their worldwide supply chain. In recent years, both managers and scholars have acknowledged the potential advantages of incorporating supply chain ethical responsibility into company performance. Commercial benefits can be derived incidentally from ethical behaviors and function as a major incentive for companies to operate ethically. Performance can be assessed using a range of measures, including both tangible and intangible factors (Bag et al. 2023).

To achieve an ethical supply chain, certain tasks need to be accomplished, such as improving operational efficiency, enhancing visibility into supplier actions, and optimizing transportation routes to minimize fuel usage. The Internet of Things (IoT), blockchain technology, artificial intelligence (AI) bots, and enhanced real-time analytics have the potential to assist in evaluating risks, optimizing workflows, reducing waste, and facilitating strategic planning. Increasingly, firms are recognizing the necessity of establishing an ethical supply chain due to consumer expectations of social responsibility in return for brand loyalty (Commerce 2022).

5.3.5.2 Top Supply Chain Companies of 2022

Each year, Gartner, Inc., releases the results of its Global Supply Chain Top 25 study, identifying the most successful supply chains and their best practices according to the highest possible environmental, social, and governance (ESG) score. The top five companies are as follows: Cisco Systems (for the third year), Schneider Electric, Colgate-Palmolive, Johnson & Johnson, and PepsiCo. According to the company, "ESG is a major focus, with circular concepts incorporated into the engineering, processes, and usage aspects of the products and supply chain. "To recognize outstanding sustained performance in the supply chain, Gartner instituted the "Masters" category in 2015. To qualify as a Master, a company must have finished in the top five in the composite score for seven of the preceding ten years. As with previous year's Masters, 2022 winners were Amazon, Apple, P&G, McDonald's, and Unilever (Gartner 2023). They and the other ranked companies can serve as examples for other chief supply chain officers (CSCOs) to follow (Table 5.1).

Following are some of the crucial factors that make the top companies successfully employ ethical practices in their SCM:

A. The role of chief supply chain officer (CSCO)

Table 5.1 Top supply chain companies of 2022

Rank	Company	Composite score
1	Cisco systems	6.71
2	Schneider electric	6.03
3	Colgate–Palmolive	5.76
4	Johnson & Johnson	5.62
5	PepsiCo	5.03
6	Pfizer	4.86
7	Intel	4.72
8	Nestle	4.70
9	Lenovo	4.60
10	Microsoft	4.58
11	L'Oréal	4.45
12	The Coca–Cola company	4.36
13	Nike	4.31
14	Walmart	4.12
15	HP Inc	3.99
16	Diago	3.95
17	Dell Technologies	3.94
18	Inditex	3.93
19	BMW	3.76
20	AbbVie	3.66
21	Siemens	3.42
22	AstraZeneca	3.40
23	General Mills	3.31
24	British American Tobacco	3.23
25	Alibaba	3.15

Source: Gartner (2023)

CSCOs are now widely recognized as influential forces behind novel economic models, green movements, and product development. To fulfill these new responsibilities, supply chain leaders are looking outside their companies to form collaborative ecosystems that can tackle more systemic issues. "For example, many of the ranking's consumer goods companies are banding together through the Consumer Goods Forum to set the agenda for developing new plastic recycling technologies." This kind of collaboration is an efficient strategy for handling the environmental issue (Marculetiu et al. 2023).

B. Stabilized supply chains

The most successful supply chains have adjusted their methods of making choices and their financial structures to make them more flexible in the face of constant change. To deal with transitory business issues, several companies have established transformation teams, understanding that their members would return to their previous roles once the situation has stabilized. This flexibility is made possible by venture capital-inspired forms of adaptive funding. These frameworks allow transformational spending to scale up in response to encouraging results from pilot projects or scale down if those projects do not pan out or if other priorities take precedence (Silva et al. 2023).

C. Progress on the overarching sustainability agenda

Emissions reductions are only one part of the "net zero" pledges made by leading supply chains to their suppliers and customers. Hence, this elaborates that large corporations "have formal programs to track the status of supplier reduction projects and quantify the resulting reductions over time." Examples of such corporations include Walmart, Microsoft, and Unilever, which are high-tech industries beginning to embrace circular economy strategies. Globally, IT giants like Cisco, Lenovo, and HP Inc. are moving away from the periodic sale of standalone pieces of hardware and toward service-based models that include the responsible collection, recycling, and disposal of obsolete technology (Richey et al. 2022).

D. Digital automation with a human focus

Immediate investments in technology that lowers employee cognitive load and prioritizes time and attention to areas where humans outperform machines, such as relationship building and adjusting to changing operational circumstances, should be balanced with long-term investments in automation. There are other instances when humans outperform robots. Warehouse or industrial employees, for example, may boost their efficiency by collaborating with Industrial Internet of Things (IIOT). Leaders are prioritizing programs that train digital literacy and dexterity to prepare their employees for the future. These educational and applied learning programs enable employees to use data-driven analytics to make more informed decisions (Gartner 2023).

5.4 Discussion and Implications

Companies globally are recognizing the necessity of integrating ethical standards into their business operations. Workplace ethical challenges can manifest in various ways, and organizations have the discretion to interpret and address them according to their judgment. Due to the variation in moral standards and norms of conduct throughout different nations, businesses lack a generally accepted standard due to which global firms encounter specific ethical dilemmas. Corporations operating globally are always looking for fresh opportunities to expand their operations. Corporate codes are defined by laws which are the guiding principles of business that all companies must follow. The interests of consumers and the community should be prioritized before those of any businesses (phDessay 2022). The literature on smart SCM, buyer–supplier interactions, and business process re-engineering has yet to address the broader topic of managing ethics in supply chain relationships. In light of these considerations, it is clear that ethical smart SCM is crucial to satisfying customers where their needs must always come first to maintain satisfaction and the reliability of the supply chain.

Ethical smart SCM aligns technological advances and operational efficiencies with human values and societal expectations through ethics. Maintaining trust and integrity when firms implement smart technologies like AI and blockchain to improve supply chain operations requires ethical apprehensions. Smart SCM integrated with ethical principles eliminates various issues such as misuse of human capital management, and environmental devastation caused by a hasty technological deployment. As ethics promotes transparency and accountability, which boosts stakeholder trust and brand reputation, it also reduces legal risks by following regulations and international guidelines. Companies can support sustainable development, human rights, and corporate social responsibility by incorporating ethical concepts into smart supply chain strategy. This ethical foundation promotes long-term commercial success and social good, showing that morality and prosperity can coexist.

5.4.1 Implications for Business Managers and Entrepreneurs

Considering the predominant self-sufficiency and self-centeredness of most young individuals, the crucial inquiry that must be addressed is how the subsequent generation will effectively handle the emerging issues within the supply system. Effective management of supply networks necessitates the flexibility to adapt and scale comprehensively. The management of the supply chain will be influenced by ethical considerations, as this factor plays a crucial role in the organization's overall success. The behavior of consumers is crucial and should be regarded as a significant aspect of the long-term sustainability of the firm. Ethical considerations surrounding consumer behavior and SCM will enhance the level of customer satisfaction, specifically about the organization's products and services (Casciani et al. 2022).

However, numerous existing and prospective suppliers may dismiss the problems as insignificant, particularly if they pertain mostly to small or medium-sized businesses. An organization's ethical guidelines have a profound influence on how employees manage customer relationships, as customer engagement directly contributes to the company's financial gains (Afolabi et al. 2023). The reason for this is that an emphasis on ethics encourages the joint development of products, the adoption of standards that apply to the entire industry, and the sharing of information regarding consumer preferences among retailers. Ethical considerations significantly influence the development of an interconnected supply chain. When ethical considerations are effectively included and integrated, they can enhance physical distribution, broaden the utilization of outsourcing and partnerships, assess employee performance, and simplify the integration of moral issues into regular business activities (Alsoussi et al. 2024).

5.4.2 Implications for Academia

According to Atta-Obeng and Dadzie (2020), higher education institutions have traditionally served as the "cradle" for the education of future decision-makers, academics, entrepreneurs, leaders, and professionals. It is possible to see universities as a developing network of ethical systems that regulate the institution's teaching, research, service, and administration. On the other hand, the university system is transforming, with new regulations, methods of operation, and new concepts being added to its toolbox of methods. Today's universities are governed by a multitude of internal and external regulations and standards that handle various themes such as cheating, research on human subjects, academic integrity, research on animals, and environmental ethics. The interconnected ecological systems of ethics did not emerge from a singular logical process. Rather, they represent the ongoing historical development of law and ethics about the creation of novel values (Bennett 2023).

Providing a complete undergraduate curriculum in the SCM industry has been an arduous task due to its extensive scope. Only a small number of students choose to pursue a bachelor's degree program with a specific focus on procurement, logistics, or operations, which are the three main areas of study in the field of SCM (Drake et al. 2023). Universities continue to modify their programs in response to industry feedback, particularly from companies that recruit on campus. However, these companies consult institutions and even send executives to teach or help develop curriculums to get the type of graduates they desire. While particular organizations may concentrate on operations or procurement, others may highlight concerns regarding ethics (Ishfaq et al. 2022).

For instance, Michigan State University (MSU) employs a balanced strategy. They focus on logistics, operations, and procurement, which, is a rarity among universities, as few offer all three equally. Other universities concentrate more on operations management or procurement, which they call SCM (Vazquez Reyes et al. 2023). Employers bring actual and fabricated cases to the classroom so that students can investigate these issues using supply chain simulation software. Alternatively, they will organize a classroom competition where students must develop solutions without all the necessary information. Veterans of the industry can learn about their career requirements and the needs of corporations by observing what universities are doing with their supply chain programs. Obtaining a degree program and receiving relevant training will enable supply chain managers to secure their initial and potentially subsequent job opportunities. (Content deleted due to incoherence). Ambitious supply chain administrators must possess a comprehensive understanding of their company's complete end-to-end supply chain, rather than solely focusing on their job responsibilities. They must know about the intricate and demanding connections between their suppliers and consumers in the current situation (Vazquez Reyes et al. 2023).

5.4.3 Implications for Policymakers

SCM experts are responsible for promoting ethical actions throughout the smart supply chain. Enhanced financial resources result in the generation of higher brand and shareholder value. In addition, the appeal of highly proficient employees promotes customer and supplier loyalty, ultimately leading to a decrease in legal disputes against firms with strong ethical values. Considering the increasing importance of ethics in smart SCM, any decisions made at any point in the chain will have a substantial impact on the overall operation (Jiang et al. 2019). The policy should promise to protect all workers' access to globally recognized protections and eliminate all kinds of child or forced labor. Workers should be granted the right to freely associate and engage in collective bargaining without any kind of discrimination. Ensuring the well-being of employees should take precedence by providing them with reasonable working hours. The policy should explicitly outline worker rights such as the prohibition of abusive practices or disproportionate disciplinary actions, protection of legal rights, provision of living wages, and equitable benefits (Ferrell and Fraedrich 2021).

Supply chains in the present day are a steadfast pillar of integrity and a dependable means to protect a company's future. Mere awareness of the activity of the closest suppliers is no longer sufficient. Businesses will face challenges in effectively monitoring their suppliers' ethical and responsible procurement practices, which will hinder their ability to oversee suppliers' overall performance (Autry 2021). To build and maintain a supply chain that is both sustainable and ethical, information is crucial, where the efficient acquisition, analysis, and use of large data sets achieve an Information advantage. In addition, supply chain openness and visibility help companies maintain ethical standards and best practices (Khan et al. 2022). The CEO of a company is responsible for keeping an eye on a wide range of issues related to its operations and those of its suppliers, customers, and other business partners. Hygienic workplaces, equitable wages and working hours, anti-bribery and anti-corruption measures, ethical sourcing and acquisition, environmental awareness and sustainability, and the elimination of child and slave labor are all central tenets of ethical smart supply chains that aim for the highest levels of ethical, sustainable and efficient operations. If a supplier is shown to be in breach of agreed-upon ethical standards, the firm may promptly take action to rectify the situation, thereby safeguarding the company's reputation. Individuals employed in supply chains that prioritize operational transparency and visibility possess a clear edge over their competitors, as they guarantee that their organization adheres to all ethical norms (Snyder 2020).

5.5 Conclusion

Ethical issues in smart SCM are intricate and multifaceted, involving various stakeholders, including suppliers, manufacturers, distributors, retailers, and consumers. Labor practices, environmental impact, human rights, and corruption are among the primary ethical concerns in SCM. To ensure that ethical standards are reached, businesses must assume responsibility for their complete supply chain, from procuring basic materials to delivering the ultimate product. This requires transparency, accountability, supplier collaboration, and continuous monitoring and enhancement of supply chain practices. The treatment of workers is a major ethical issue in the SCM industry. Numerous businesses rely on low-cost labor in developing nations, where employees may be subjected to poor working conditions, low wages, and even forced labor. In many countries, these practices are not only reprehensible but also prohibited. Companies are responsible for ensuring that their suppliers comply with labor laws and international labor standards, such as the fundamental labor standards of the International Labor Organization (ILO). Additionally, human rights and corruption are significant concerns in SCM, especially in developing nations. Bribery and other forms of corruption can undermine ethical standards and lead to unequal competition, market distortions, and monetary losses. Companies must employ a zero-tolerance corruption policy and implement measures by using smart supply chain technologies to detect and prevent corrupt practices in their supply chains.

In summary, it is crucial to uphold ethical values in smart SCM to promote a sustainable and socially responsible company environment. Adhering to ethical principles guarantees that the incorporation of cutting-edge technologies such as AI and blockchain improves operational effectiveness while upholding human rights, environmental sustainability, and stakeholder trust. By giving priority to ethics, companies can avoid engaging in exploitative behaviors, improve transparency, and establish a strong reputation. This ethical commitment not only mitigates legal and regulatory risks but also conforms to the increasing consumer and investor demand for conscientious company conduct. Incorporating ethical principles into smart SCM ultimately establishes a harmonious strategy that fosters sustainable profitability while making a good impact on society and the environment.

References

Achillles. 2022. *The challenges of building ethical supply chains on a global scale* [Online]. Website: Achillles. Available: https://www.achilles.com/industry-insights/the-challenges-of-building-ethical-supply-chains-on-a-global-scale/ [Accessed 23 march 2023].

ADP. 2020. *Ethical Issues in Supply Chain Analytical Essay* [Online]. Website: IvyPanda. Available: https://ivypanda.com/essays/ethical-issues-in-supply-chain/ [Accessed 28-February 2023].

ADP. 2022. *8 ethical issues in business and how to manage them* [Online]. Website: Indeed. Available: https://uk.indeed.com/career-advice/career-development/ethical-issues-in-business [Accessed 27-February 2023].

AFOLABI, J. O. A., OLATOYE, F. O., EBOIGBE, E. O., ABDUL, A. A. & DARAOJIMBA, H. O. 2023. Revolutionizing retail: hr tactics for improved employee and customer engagement. *International Journal of Applied Research in Social Sciences,* 5, 487-514.

ALAM, S. H., JIANG, Y., ZAMAN, S. I., JALEES, T. & AHMED, R. R. 2023. Behavioral biases in consumer paradox towards counterfeit luxury products: a comparative study between China and Pakistan. *Current Psychology*, 1–16.

ALSOUSSI, A., SHBIKAT, N., ALMAKHAMREH, S. & ALLAN, B. 2024. Toward more ethically oriented humanitarian logistics operations: An exploratory research. *International Journal of Disaster Risk Reduction,* 106, 104418.

ATTA-OBENG, L. & DADZIE, P. S. 2020. Promoting sustainable development goal 4: The role of academic libraries in Ghana. *International information & library review,* 52, 177-192.

AUTRY, C. W. 2021. Supply chain research: considering the discipline's evolving relationship with marketing, current issues, and future research directions. *Journal of Marketing Theory and Practice,* 29, 101-113.

AYTAN, Y. S. & SAYAN, I. 2019. Ethics, Ethical Leadership, and Supply Chain Management. *Ethical and Sustainable Supply Chain Management in a Global Context.* IGI Global.

BAG, S., RAHMAN, M. S., SRIVASTAVA, G., SHORE, A. & RAM, P. 2023. Examining the role of virtue ethics and big data in enhancing viable, sustainable, and digital supply chain performance. *Technological Forecasting and Social Change,* 186, 122154.

BENNETT, J. 2023. Static in Process: A Key to Applying Process Philosophy for Ecological Civilization. *Process Studies,* 52, 64-94.

BURTLER, J. 2022. *Ethical Issues in Business* [Online]. Website: Study.com. Available: https://study.com/learn/lesson/ethical-issues-in-business-types-handling-examples.html [Accessed 27-February 2023].

CARROLL, A. B. & BROWN, J. 2022. *Business & society: Ethics, sustainability & stakeholder management*, Cengage Learning.

CASCIANI, D., CHKANIKOVA, O. & PAL, R. 2022. Exploring the nature of digital transformation in the fashion industry: opportunities for supply chains, business models, and sustainability-oriented innovations. *Sustainability: Science, Practice and Policy,* 18, 773-795.

CHARLES, V., EMROUZNEJAD, A. & GHERMAN, T. 2023. A critical analysis of the integration of blockchain and artificial intelligence for supply chain. *Annals of Operations Research,* 1–41.

CNBC, S., N., 2023. 2023. *McDonald's franchisees fined after 305 minors, including 10-year-olds, found working illegally* [Online]. Available: https://www.cnbc.com/2023/05/03/mcdonalds-franchisees-fined-over-child-labor-law-violations.html [Accessed 2023].

COMMERCE, T. F. O. 2022. *The ethical supply chain: Definition, examples, stats* [Online]. Website: The Future of Commerce Available: https://www.the-future-of-commerce.com/2020/01/22/ethical-supply-chain-definition-stats/ [Accessed 24 march 2023].

DAY, D. V., RIGGIO, R. E., TAN, S. J. & CONGER, J. A. 2021. Advancing the science of 21st-century leadership development: Theory, research, and practice. Elsevier.

DRAKE, M. J., PITTMAN, P. H. & TALBERT, M. 2023. Analysis of the incorporation of current trends in supply chain management within curricula. *International Journal of Logistics Research and Applications*, 1–21.

FERRELL, O. C. & FRAEDRICH, J. 2021. *Business ethics: Ethical decision making and cases*, Cengage learning.

GARTNER. 2023. *Gartner Announces Rankings of the 2022 Global Supply Chain Top 25 [WWW Document]* [Online]. Available: https://www.gartner.com/en/newsroom/press-releases/2022-05-26-gartner-announces-rankings-of-the-2022-global-supply-chain-top-25 [Accessed 2023/4/30].

GILPIN, J. 2022. *Understanding Ethical Issues Within the Global Supply Chain* [Online]. Website: Supply2Gov 2023. Available: https://www.supply2govtenders.co.uk/ethical-issues-with-global-supply-chain/ [Accessed 23-February 2023].

HAZEN, B. T., RUSSO, I., CONFENTE, I. & PELLATHY, D. 2021. Supply chain management for circular economy: conceptual framework and research agenda. *The International Journal of Logistics Management*, 32, 510-537.

HOLGADO, M. & NIESS, A. 2023. Resilience in global supply chains: Analysis of responses, recovery actions and strategic changes triggered by major disruptions. *Supply Chain Management: An International Journal*, 28, 1040-1059.

HOLTON, L. 2020. *8 Ethical Issues Businesses Face Today* [Online]. Website: myVA360. Available: https://myva360.com/blog/8-ethical-issues-business-face-today [Accessed 27-February 2023].

HOUSE, D. 2022. *Why Ethics in Supply Chain Management Must Be a Priority* [Online]. Website: Shyft Available: https://www.shyftservices.com/resources/ethics-in-supply-chain-management [Accessed 28-February 2023].

ISHFAQ, R., DAVIS-SRAMEK, B. & GIBSON, B. 2022. Digital supply chains in omnichannel retail: A conceptual framework. *Journal of Business Logistics*, 43, 169-188.

JIANG, Y., MIAO, M., JALEES, T. & ZAMAN, S. I. 2019. Analysis of the moral mechanism to purchase counterfeit luxury goods: evidence from China. *Asia Pacific Journal of Marketing and Logistics*, 31, 647-669.

JIANG, Y., MUBARIK, M. S., ZAMAN, S. I., ALAM, S. H. & ARIF, M. 2023. Corporate governance, cash holding, and firm performance in an emerging market. *International Journal of Finance & Economics*.

KHAN, S. A., MUBARIK, M. S., KUSI-SARPONG, S., GUPTA, H., ZAMAN, S. I. & MUBARIK, M. 2022. Blockchain technologies as enablers of supply chain mapping for sustainable supply chains. *Business Strategy and the Environment*, 31, 3742-3756.

KOUHIZADEH, M., SABERI, S. & SARKIS, J. 2021. Blockchain technology and the sustainable supply chain: Theoretically exploring adoption barriers. *International journal of production economics*, 231, 107831.

KROSEL, A. 2022. *The Importance of Ethics in the Workplace: 6 Significant Benefits* [Online]. Website: careerguideindeed. Available: https://www.bing.com/ck/a?!&&p=3c6d5a50d82e89f3JmltdHM9MTY3OTcwMjQwMCZpZ3VpZD0zNmUxNzNmOS02OGZlLTZjNGItMmQ4Zi02MTI3NjkyNTZkYmQmaW5zaWQ9NTI3Ng&ptn=3&hsh=3&fclid=36e173f9-68fe-6c4b-2d8f-612769256dbd&psq=what+is+ethical+issues+in+workplace&u=a1aHR0cHM6Ly93d3cuaW5kZWVkLmNvbS9jYXJlZXItYWR2aWNlL2NhcmVlci1kZXZlbG9wbWVudC93aHktZXRoaWNzLWlzLWltcG9ydGFudC1pbi10aGUtd29ya3BsYWNl&ntb=1 [Accessed 25 march 2023].

KUNDUZOVIHC, M. M. 2022. Emerging technologies in supply chain management.

KWARTENG, A., SIMPSON, S. N. Y. & AGYENIM-BOATENG, C. 2022. The effects of circular economy initiative implementation on business performance: the moderating role of organizational culture. *Social Responsibility Journal*, 18, 1311-1341.

LEBARON, G. 2021. The role of supply chains in the global business of forced labour. *Journal of Supply Chain Management*, 57, 29-42.

MANNING, L., BREWER, S., CRAIGON, P. J., FREY, J., GUTIERREZ, A., JACOBS, N., KANZA, S., MUNDAY, S., SACKS, J. & PEARSON, S. 2022. Artificial intelligence and ethics within the food sector: Developing a common language for technology adoption across the supply chain. *Trends in Food Science & Technology*.

MARCULETIU, A., ATASEVEN, C. & MACKELPRANG, A. W. 2023. A review of how pressures and their sources drive sustainable supply chain management practices. *Journal of Business Logistics,* 44, 257-288.

MARTIN, D. A., CONLON, E. & BOWE, B. 2021. Using case studies in engineering ethics education: The case for immersive scenarios through stakeholder engagement and real life data. *Australasian Journal of Engineering Education,* 26, 47-63.

MARTINS, K. F., TEIXEIRA, D. & BATALH O, A. C. 2023. Voluntary Sustainability Standards for Corporate Social Responsibility. *The Route Towards Global Sustainability: Challenges and Management Practices.* Springer.

MC LOUGHLIN, K., LEWIS, K., LASCELLES, D. & NUDURUPATI, S. 2023. Sustainability in supply chains: reappraising business process management. *Production Planning & Control,* 34, 19-52.

MELLA, P. 2019. Quality, a Key Value Driver in Value Based Management. *Economia Aziendale Online-,* 9, 439-462.

MISHRA, A., BADHOTIYA, G. K., PATIL, A., SIDDH, M. M. & RAM, M. 2023. Servitization in the circular supply chain: delineating current research and setting future research plan. *Management of Environmental Quality: An International Journal,* 34, 1035-1056.

MUBARIK, M. S., KAZMI, S. H. A. & ZAMAN, S. I. 2021. Application of gray DEMATEL-ANP in green-strategic sourcing. *Technology in Society,* 64, 101524.

NAVEED, R. T., ALHAIDAN, H., AL HALBUSI, H. & AL-SWIDI, A. K. 2022. Do organizations really evolve? The critical link between organizational culture and organizational innovation toward organizational effectiveness: Pivotal role of organizational resistance. *Journal of Innovation & Knowledge,* 7, 100178.

NILSSON, F. & G RANSSON, M. 2021. Critical factors for the realization of sustainable supply chain innovations-model development based on a systematic literature review. *Journal of Cleaner Production,* 296, 126471.

NOMI, M. 2022. *The Importance of Supply Chain Ethics and Compliance and Top 6 Best Practice Tips for Every Company* [Online]. Website: polonious. Available: https://www.polonious-systems.com/supply_chain_ethics_and_compliance/ [Accessed March 10 2023].

NUSEIR, M. T. & GHANDOUR, A. 2019. Ethical issues in modern business management. *International Journal of Procurement Management,* 12, 592-605.

PhDessay. 2022. *Ethical Issues Faced by Global Organizations* [Online]. Website: phDessay. Available: https://www.bing.com/ck/a?!&&p=9d02e43ca2e006e0JmltdHM9MTY4MDQ4MDAwMCZpZ3VpZD0yYzA1YTViMy00YTdkLTZhMjktMjU4MC1iNzU0NGI3NDZiNWMmaW5zaWQ9NTE4Mw&ptn=3&hsh=3&fclid=2c05a5b3-4a7d-6a29-2580-b7544b746b5c&psq=Ethical+Issues+Faced+by+Global+Organizations&u=a1aHR0cHM6Ly9waGRlc3NheS5jb20vZXRoaWNhbC1pc3N1ZXMtZmFjZWQtYnktZ2xvYmFsLW9yZ2FuaXphdGlvbnMv&ntb=1 [Accessed 3 April 2023].

RADANLIEV, P., DE ROURE, D., ANI, U. & CARVALHO, G. 2021. The ethics of shared Covid-19 risks: an epistemological framework for ethical health technology assessment of risk in vaccine supply chain infrastructures. *Health and Technology,* 11, 1083-1091.

RICHEY, R. G., ROATH, A. S., ADAMS, F. G. & WIELAND, A. 2022. A responsiveness view of logistics and supply chain management. *Journal of Business Logistics,* 43, 62-91.

ROGERSON, M. & PARRY, G. C. 2020. Blockchain: case studies in food supply chain visibility. *Supply Chain Management: An International Journal,* 25, 601-614.

ROUNAGHI, M. M., JARRAR, H. & DANA, L.-P. 2021. Implementation of strategic cost management in manufacturing companies: overcoming costs stickiness and increasing corporate sustainability. *Future Business Journal,* 7, 1-8.

SCHROEDER, K. 2021. *3 Reasons Why Business Ethics Is Important* [Online]. Website: redlands. Available: https://www.redlands.edu/study/schools-and-centers/business/sbblog/2019/may-2019/3-reasons-why-business-ethics-important/ [Accessed 10-march 2023].

SETTEMBRE-BLUNDO, D., GONZ LEZ-S NCHEZ, R., MEDINA-SALGADO, S. & GARC A-MUI A, F. E. 2021. Flexibility and resilience in corporate decision making: a new sustainability-based risk management system in uncertain times. *Global Journal of Flexible Systems Management,* 22, 107-132.

SHAUN. 2022. *What are Ethical Issues?* [Online]. Website: MyAccountingCourse.com Available: https://www.myaccountingcourse.com/ [Accessed 27-February 2023].

SHETTY, S. K. & BHAT, K. S. 2022. Green supply chain management practices implementation and sustainability–A review. *Materials Today: Proceedings,* 52, 735-740.

SILVA, M. E., PEREIRA, M. M. & HENDRY, L. C. 2023. Embracing change in tandem: Resilience and sustainability together transforming supply chains. *International Journal of Operations & Production Management,* 43, 166-196.

SNYDER, J. 2020. *How ethical & sustainability supply chain governance can help maintain the planet* [Online]. Website: CIO Executive Editorial. Available: https://www.cio.com/article/201921/how-ethical-sustainability-supply-chain-governance-can-help-maintain-the-planet.html [Accessed 19 April 2023].

STAFF, S. 2019. *Why Business Leaders Must Make Supply Chain Ethics a Priority* [Online]. Website: Paul Kenny. Available: https://www.scmr.com/article/why_business_leaders_must_make_supply_chain_ethics_a_priority [Accessed 30 March 2023].

TAIYABMUSTAFIZ. 2022. *Ethical Issues* [Online]. Website: geeksforgeeks. Available: https://www.geeksforgeeks.org/human-values-concept-and-importance/?ref=leftbar-rightbar [Accessed 27-february-2023 2023].

TSAI, F. M., BUI, T. D., TSENG, M.-L., CHIU, A. S. & CHIU, S. 2021. *Sustainable supply chain management trends in world regions: A data-driven analysis* [Online]. Website: researchgate. Available: https://www.researchgate.net/publication/350559444_Sustainable_supply_chain_management_trends_in_world_regions_A_data-driven_analysis [Accessed 27 April 2023].

TUPS, G. & DANNENBERG, P. 2023. Supplying lead firms, intangible assets and power in global value chains: explaining governance in the fertilizer chain. *Global Networks,* 23(4), 772–791.

TWIN, A. 2022. *Business Ethics: Definition, Principles, Why They're Important* [Online]. Website: Investopedia. Available: https://www.investopedia.com/terms/b/business-ethics.asp [Accessed 27-February 2023].

UCARYILMAZ DEIBEL, T. 2022. Corporate social responsibility in the legal framework of global value chains. *Law and Development Review,* 15, 329-356.

ULLAH, S., ADAMS, K., ADAMS, D. & ATTAH-BOAKYE, R. 2021. Multinational corporations and human rights violations in emerging economies: Does commitment to social and environmental responsibility matter? *Journal of Environmental Management,* 280, 111689.

DI VAIO, A., VARRIALE, L., LEKAKOU, M. & POZZOLI, M. 2023. SDGs disclosure: evidence from cruise corporations' sustainability reporting. *Corporate Governance: The International Journal of Business in Society.*

VAZQUEZ REYES, B. O., TEIXEIRA, T., COLMENERO, J. C. & PICININ, C. T. 2023. Assessing educational methods for tomorrow's supply chain leaders with the integration of skill development priorities: a fuzzy decision-making approach. *Journal of Enterprise Information Management,* 36, 349-380.

VITOLLA, F., RAIMO, N., RUBINO, M. & GAREGNANI, G. M. 2021. Do cultural differences impact ethical issues? Exploring the relationship between national culture and quality of code of ethics. *Journal of International Management,* 27, 100823.

WAINWRIGHT, L. & MU OZ, P. 2020. Restorative entrepreneuring: A new cross-disciplinary agenda to support at-risk social groups. *Journal of Business Venturing Insights,* 13, e00165.

WATERS, D. 2021. *Logistics An Introduction to supply chain management*, Palgrave macmillan.

YANG, Y. 2022. *Digital Economy and the Sustainable Development of ASEAN and China*, World Scientific.

YUSUF, Y., HAWKINS, A., MUSA, A., EL-BERISHY, N., SCHULZE, M. & ABUBAKAR, T. 2014. Ethical supply chains: analysis, practices and performance measures. *International Journal of Logistics Systems and Management,* 17, 472-497.

Zaman, S. A. A., Dawood, H. M., Zehra, S. N. & Saeed, S. Z. 2023a. Blockchain Driven Supply Chain and Industry 4.0 Technologies. *In:* MUBARIK, M. S. & SHAHBAZ, M. (eds.) *Blockchain Driven Supply Chain Management: A Multi-dimensional Perspective.* Singapore: Springer Nature Singapore.

ZAMAN, S. I., KHAN, S., ZAMAN, S. A. A. & KHAN, S. A. 2023b. A grey decision-making trial and evaluation laboratory model for digital warehouse management in supply chain networks. *Decision Analytics Journal*, 100293.

ZHANG, G., YANG, Y. & YANG, G. 2023. Smart supply chain management in Industry 4.0: the review, research agenda and strategies in North America. *Annals of Operations Research,* 322, 1075-1117.

Chapter 6
Complexities in Smart Supply Chain and Role of Contemporary Leadership

Arooj Rafiq, Shamim Akhtar, Masood Nawaz Kalyar, Agata Pierscieniak, and Denisa Bogdana Abrudan

6.1 Introduction

In the current era of rapid technological advancements, digital transformation (DT) has garnered significant attention from academia and industry due to its profound impact across sectors and aspects of human life. The intersection of leadership and digital technologies has gained unprecedented importance as industries increasingly integrate sophisticated tools like AI, IoT, and blockchain. This integration has redefined supply chain operations, enhancing efficiency, transparency, and responsiveness, yet it also introduces complex challenges that require adept and visionary leadership. In the context of the smart supply chain, effective leadership is crucial at every stage: planning, adoption, and post-adoption. Leaders are responsible for setting strategic visions, guiding the integration of these advanced technologies, and ensuring ongoing improvement post-implementation. They must balance innovation with practical execution, fostering a culture that embraces technological advancements while maintaining operational efficiency. Contemporary leaders in smart supply chains are thus tasked with not only navigating these complexities but also creating agile, collaborative environments that drive resilience, adaptability, and

A. Rafiq · S. Akhtar
Department of Business Administration, Faculty of Management Sciences, University of Sialkot, Sialkot, Pakistan

M. N. Kalyar (✉)
Lyallpur Business School, Government College University Faisalabad, Faisalabad, Pakistan
e-mail: masood.kalyar@yahoo.com

A. Pierscieniak
Faculty of Management, Warsaw University of Technology, Warsaw, Poland

D. B. Abrudan
Faculty of Economics and Business Administration, West University of Timisoara, Timisoara, Romania

M. S. Mubarik and S. A. Khan (eds.), *Smart Supply Chain Management*,
https://doi.org/10.1007/978-981-96-1333-5_6

sustained performance in a rapidly evolving digital landscape (Margherita et al. 2025; Chatterjee et al. 2023).

The intersection of digital transformation and social change presents a complex landscape for contemporary business leaders, particularly within smart supply chains. While there has been significant discussion on the benefits and implementation strategies of smart supply chains, the specific role of leadership in this context remains underexplored. Existing literature often overlooks how leadership directly influences the effectiveness and long-term sustainability of smart supply chain initiatives. This gap underscores the need for a focused examination of leadership's impact on smart supply chain management. In this context, leaders play a critical role not only in facilitating the integration of advanced technologies like AI, IoT, and blockchain but also in aligning these changes with broader social and ethical considerations. Leaders are responsible for navigating each phase of the smart supply chain lifecycle—planning, adoption, and post-adoption—by setting strategic directions, managing the intricacies of technology adoption, and ensuring continuous improvement post-implementation. Their role involves balancing innovation with practical execution, fostering a culture that embraces technological advancements while being attentive to societal shifts and sustainability (Buonocore et al. 2024; Bastidas et al. 2023).

The objectives of this chapter are to examine these critical roles of leadership within smart supply chains and provide insights into how leaders can effectively manage and optimize these technologies to enhance overall supply chain performance. Additionally, this chapter aims to identify leadership strategies that drive successful smart supply chain initiatives, offering practical guidance for businesses and highlighting areas for future research at the intersection of leadership and supply chain technology.

The structure of the chapter is as follows: it begins with an exploration of the motivation behind studying the role of leadership in smart supply chains. This is followed by an overview of smart supply chain concepts and the technological tools involved. The relationship between AI and SSCs is then examined, including the benefits and challenges of AI integration. Subsequent sections discuss the enablers and barriers to AI adoption, providing a comprehensive view of the factors influencing successful implementation. The chapter concludes with discussions on future research directions and practical implications for industry professionals and policymakers.

6.2 Leadership and Its Role in SCM

Leadership, defined as the capacity to influence and guide individuals or groups toward a specific goal or vision (Rees and Porter 2002), plays a pivotal role in smart supply chain management (SCM). In SCM, leadership extends beyond individual influence to encompass the strategic ability to coordinate activities across functions and firms. Effective leadership is essential for aligning supply chain operations with organizational goals, driving synchronization, and maintaining focus amid complex, technology-driven processes. In the context of smart supply chains, leadership positioning—whether a Tier 1 or Tier 2 supplier takes the lead—significantly impacts CSR efforts and violation risks. When Tier 2 suppliers lead, delegation strategies can reduce violation probabilities; conversely, a leading Tier 1 supplier may reduce CSR efforts to cut costs, increasing risks if a uniform funding model is applied. Adjusting the funding factor to be endogenous provides buyers with flexibility, often favoring Tier 1 leadership for better compliance control at Tier 2. Under stringent CSR investment constraints, the focus shifts to suppliers with cost advantages, independent of hierarchical position, reflecting practices seen in industry leaders like Apple. This underscores that contemporary leadership in SCM is not only about driving operational efficiency but also about balancing ethical and CSR responsibilities, navigating complex funding and delegation strategies to foster resilient and socially responsible supply chain ecosystems. Through effective leadership, organizations can align smart supply chain activities with broader goals, enhancing both performance and sustainability (Liu et al. 2022; Cannon-Bowers et al. 1991).

Managers with a good SCM background can effectively foster the successful management of culture throughout the supply chain, resulting in more effective and quality-based decisions, efficient supply chain responsiveness, and enhanced customer satisfaction. Some of the leadership competencies that are embodied in SCM include communication, decision-making, risk management, and change management. The other attribute within SCM is that a strong leader in SCM must also be willing and capable of fostering and sustaining good working relationships with the supply chain stakeholders such as the suppliers and customers (Salas et al. 2012). They are in a position to harness the opportunities presented by technology and data analytics to enhance the physical and virtual supply chain visibility and efficiency as well as cut costs. In conclusion, it was found that leadership had a significant influence on the future of supply chain management. Management competence is critical in attaining operational efficiency, enhancing customer value, and good economic performance. The leadership and smart supply chain management gap is the deficiency in the linkage between effective leadership of organizations and the implementation of best practices in the provision of smart supply chain solutions that are informed by information technology. This raises awareness of situations where the leadership does not offer complete support for the promotion of smart supply chain strategies or when smart supply chain initiatives are not aligned with the organization's overarching strategic plan. It is for this reason that it is necessary to look at possible causes and embrace research as the primary method of filling this gap.

Resistance to change is therefore notable as one of the key barriers to realizing smart supply chain initiatives in an organization. This can lead to employees and management not supporting innovative technologies and strategies that are important in supply chain performance. To overcome this resistance, there is a need for more research to be conducted in the area of leadership and change to understand how to counter opposition. Ignorance and unawareness can be another barrier to the supply chain management adoption among executives. Managers may simply not be aware that smart supply chain management can be advantageous for their business and helpful in the context of competition. Furthermore, they may easily be out of touch with the existing supply chain technology advancements and their implications for their firm. To fill such knowledge gaps and ensure the public is aware of the new developments within the industry is the role of research.

To achieve the framework of smart supply chain projects, strategic alignment is an essential ingredient. These efforts may not align with the organizational strategic plan if they are not backed by leadership support and guidance and end up being costly for the organization with little possibility of optimization. Research should therefore stress the importance of strategy in the smart supply chain initiative by the organization. Organizational Silos disrupt the proper flow of information and knowledge for effective supply chain management. When multiple teams need to be aligned and work simultaneously on achieving multiple organizational goals and objectives, effective leadership is critical. They should further investigate ways to overcome these barriers and encourage integration of the supply chain management approach. A lack of appropriate data-driven culture within an organization may act as a constraint to the actualization of smart supply chain practices. Supply chain management is a data-driven process, and the success of demand management is no exception. Future studies must examine the need for people to embrace data-informed decision-making as a culture to support smart supply chains.

A lack of available skills and talent can hamper the development of smart supply chain management solutions. There is a need for specialized skills where the integration process is concerned, such as data analysts, experts in artificial intelligence, and supply chain personnel who are conversant with matters technological. This challenge though can be alleviated with leadership's support to invest in the top personnel. There is a significant leadership and smart supply chain management weakness that needs to be addressed to ensure the success, sustainability, and competitiveness of organizations in the complex and growing environment. This means that with efforts being made to understand the factors that explain the existence of the gap, then the right practices of supply chain management are likely to be put in place that will enhance the strategic directions of an organization. This research underscores the need for scholarly investigation and management of this gap to facilitate effective functioning and adaptation to challenges in a complex and increasingly competitive global environment.

6.3 Leadership and Smart Supply Chain Management

Effective leadership is essential for the success of smart supply chain management (SCM). Figure 6.1 in following exhibits the keyways in which leadership impacts smart SCM:

Fig. 6.1 Impact of leadership on smart supply chain management. *Source* Authors' own processing based on literature review

6.3.1 Strategic Direction and Vision

The role of vision and strategy as leadership variables cannot be over-emphasized because they play a very central role in shaping the destiny of any organization. About SCM, vision and strategy point to the overall, often long-term, goals, aims, and process direction of the supply chains of an organization. This is how vision and strategy come into the picture (Fig. 6.2) to influence smart supply chain management.

6.3.1.1 Clear Direction

A clear direction is defined as a leader's capacity to prescribe clear and specific ways for an organization to pursue to satisfy its objectives. It refers to the ability to outline strategic direction in a supply chain context, including the determination of a long-term vision and goals, along with a plan to achieve those goals. The following are the ways clear direction influences smart supply chain management:

Focus: Coordination translates to the supply chain managers having a clear direction for their goals and objectives. With the understanding of the supply chain strategic vision and direction as well as the road map identifying achievable goals, the supply chain managers would then be in a position to direct their activities and resources where they are most needed (Siagian et al. 2020).

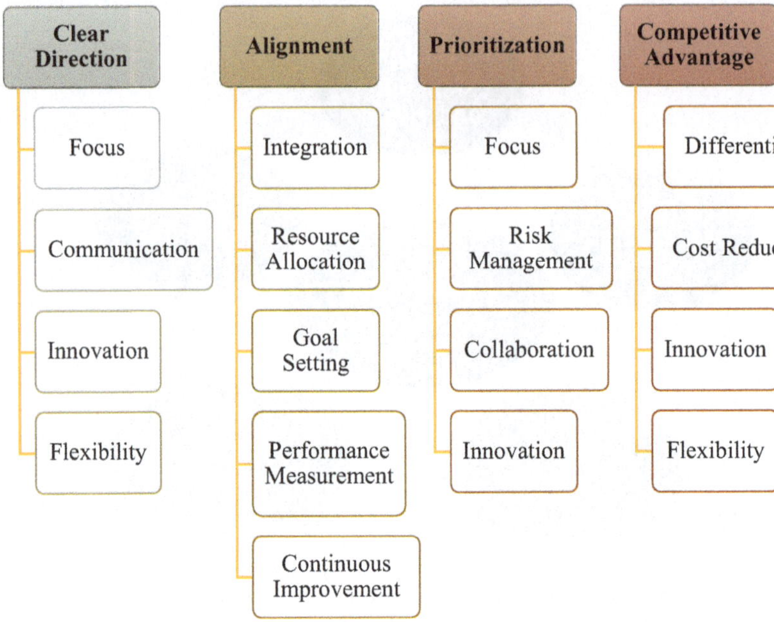

Fig. 6.2 Strategic alignment in smart supply chain management. *Source* Authors' own processing based on literature review

Communication: Clear direction enables effective communication among team members and stakeholders. By providing a clear roadmap and objectives, leaders can communicate effectively with their teams, suppliers, customers, and other stakeholders, which helps everyone understand the organization's goals and objectives (Saputra and Riana 2021).

Innovation: Clear direction fosters innovation and creativity in supply chain management. By having a clear direction, leaders can encourage their teams to come up with new and innovative ideas to achieve their goals (Gainey and Klaas 2003).

Flexibility: Lack of control is also facilitated by clear direction. A direction is important but at the same time, one has to be willing to be flexible and change direction if necessary. When the direction is clear, decisions and changes of strategy for supply chain managers become much easier for them to make (Carrell 2007; Shafique et al. 2023).

Alignment

Consistency means the act of ensuring that the strategy for the supply chain is well-coordinated with the general strategic plan of the firm. Coordination is crucial in a smart supply chain system because it allows the supply chain to continuously be tailored to meet business goals as well as enable the goals to be achieved effectively. Here are the ways alignment affects smart supply chain management:

Integration: Alignment enables coordination in the supply chain and its integration into the general business strategy. Thus aligning the supply chain strategy with business strategy will enable the supply chain managers to understand how the supply chain is supporting the achievement of organizational goals and objectives (Bibi et al. 2017).

Resource Allocation: This leads to proper coordination of the resources in that; alignment assists in the right deployment of resources for the supply chain strategy. Implementing the supply chain strategy in line with the business strategy will help the supply chain managers justify the resources required to support the strategies on the supply chain.

Goal Setting: Proper positioning assists in setting the right goals and objectives for the supply chain. That is why it is also important for the supply chain strategy to be aligned with the business strategy since this is what will define the goals and objectives of the supply chain (Asfaw et al. 2015).

Performance Measurement: The two help in the measurement of supply chain performance. This will ensure superior supply chain performance by developing a metric that fits the level of strategic alignment between the supply chain strategy and the business strategy (Bibi et al. 2017).

Continuous Improvement: Alignment also fosters ongoing improvement initiatives in the supply chain. In this respect, the supply chain strategy can be aligned with the business strategy, and subsequent improvement, initiatives can be brought in line with the goals and objectives of the organization (Terera and Ngirande 2014).

Prioritization

This pertains to the identification of the most significant and critical activities, projects, and assets in terms of their order of priority. In supply chain management leveraging smart technologies, organizations need to prioritize supply chain decisions to meet business goals and objectives. Here are some ways in which prioritization influences smart supply chain management:

Focus: Prioritization comes in handy, in that it enables the supply chain managers to allocate their efforts and resources to the pertinent initiatives. Supply chain managers must prioritize initiatives to address the highest priority initiatives first to determine the best allocation of resources.

Risk Management: Risk prioritization assists in managing risks well in a supply chain. This means that by working with the Triple Bottom Line approach, supply chain managers can determine which initiatives matter most and which ones are most likely to bring good results, therefore managing potential risks and challenges in the supply chain area before they turn into major problems (Terera and Ngirande 2014; Shafique et al. 2022).

Collaboration: A few benefits realized from prioritization include: assigning priorities to linear initiatives, SC managers can with the help of suppliers customers, and other associated players develop the desired outcomes of the supply chain (Haider et al. 2015).

Innovation: Appropriateness enhances creativity and innovation within the supply chain management. Utilizing risk and opportunity to prioritize initiatives, supply chain managers can motivate their team to think of creative ways to be able to achieve their goals (Murtiningsih 2020).

Competitive Advantage

Competitor analysis is the identification of the strengths and weaknesses of the competitors to derive a competitive advantage over them. Within the context of smart supply chain management, competition is a critical factor, which can lead a company to success, and remain relevant in the constantly growing market environment. This is how competitive advantage affects smart supply chain management:

Differentiation: Competitive advantage allows SCM professionals to position the organization strategically to other entities in the supply chain. By developing and applying special skills and opportunities inherent to the supply chain concept, supply chain managers can provide customers with some forms of value propositions that are far from easy for competitors to imitate (Bibi et al. 2017).

Cost Reduction: Sustainability in distribution networks also brings down costs within the particular chain. Due to leveraging of leveraging, special strengths, supply chain processes, and capabilities, the managers can be in effect of minimizing costs and make greater effectiveness which can lead to a competitive edge in the market (Mondol 2021).

Innovation: Business competitiveness challenges supply chain management to embrace creativity and innovation. Through applications of these specific competencies, supply chain managers can motivate the individuals in their teams to look for other opportunities and means of effectively delivering what is expected of them.

Flexibility: Competitive advantage means the supply chain managers can react to market conditions at the right time. Supply chain managers are also in a position to present a form of flexibility because they can also tap into their strengths and capacity to fashion different strategies and changes that can fit current market dynamics (Horwitz 1999).

6.3.2 *Data Analytics*

Supply chain operation information technology deals with the use of data and analytical tools to collect, process, and examine data associated with supply chain functions to make informed decisions (Chatterjee and Chaudhuri 2021). Data analytics is critical in the context of SSC management and serves as a key enabler for achieving business goals, optimizing operations, and enhancing flexibility. Here are some approaches through which data analytics affects smart supply chain management:

Demand Forecasting: It is imperative to stress that demand forecasting remains a significant part of data analytics responsible for the efficiency of smart supply chain management; leadership is also involved in this process. By using data analytic tools and data analysis techniques, an organization can gather data from multiple sources and forecast the demand levels with great precision (Frederico et al. 2020). Specifically, in the context of supply chain management, data analysis enables also accurate demand planning, based on consumption patterns observed over time. This means that through a thorough planning and forecasting of the demand of the products or services to be supplied, supply chain managers can be in a position to provide the right amount of stocks and resources required by the customers or end-users.

Cost Reduction: The supply chain management application of data analytics is useful in an endeavor to cut unnecessary expenses as it pinpoints the various areas where the flow of expenses can be restrained (Kossivi et al. 2016). Moreover, through applied and big data analytics tools, the supply chain managers can enhance, design, and manage the supply chain processes and resources, minimizing wastage and inefficiency of resources.

6.3.2.1 Change Management

Change management in supply chain management is the act of anticipating, evaluating, executing, and monitoring change in the supply management system to achieve the business objectives of a firm (Cron et al. 2005). In the aspect of SC management for a smart supply chain, change management is crucial when introducing changes to the value system, technology, and environment. The following are the ways change management affects smart supply chain management:

Planning: Strategizing is a vital strategy element in managing change in supply chain management. This mainly entails the evaluation of the existing status of supply chain management with a view to determining the potential success factors for digital transformation and the accompanying goals and objectives. Below are some of the facets of planning when it comes to change management:

Analysis: Change management planning involves assessment of supply chain functions as a basis for transformational action. This includes the use of data on inventory stock, order cycle time, delivery time, and other KPIs in the process of making suggestions for enhancement.

Setting Goals: Once the analysis is complete, the next step is to set clear goals and objectives for digital transformation (Cron et al. 2005). This involves defining the scope of the project, setting targets for KPIs, and determining a timeline for implementation.

Stakeholder Engagement: Planning in change management requires engagement with stakeholders, including employees, customers, and suppliers (Bag et al. 2021). This involves communicating the need for change, the benefits of digital transformation, and the impact on stakeholders, and addressing any concerns or questions they may have.

Resource Allocation: Implementation in change management entails resource deployment, such as resource money, people, and technology. This includes identifying digital resources that are necessary for change and ensuring that such resources are accessible and proper provisions are made for getting them (Hersey 1996).

Risk Assessment: Planning in change management requires assessment of risks and challenges that may arise during the implementation of digital transformation (Evans and Foster 2000). This involves identifying potential risks, such as disruption to supply chain operations or resistance from stakeholders, and developing strategies to mitigate these risks.

6.3.2.2 Communication

Effective communication is one of the most important factors in managing change as relates to supply chain management. This can only be developed through engagement, relation-building, and dialogue that would support the intended change. Here are some important facets concerning communication in change management:

Clear Messaging: Effective communication in change management requires clear and concise messaging (Campbell 1971). This involves clearly articulating the need for change, the benefits of digital transformation, and the impact on stakeholders, in a way that is easy to understand.

Consistent Messaging: Stakeholders should exchange information on change management consistently. This involves making sure that the message sent to all people is similar in content irrespective of the capacity of the person or involvement of the person in the project at hand.

Two-Way Communication: It is important to note that for communications in change management to be effective, they have to be two-way. This includes involving all the stakeholders in the decision-making process, listening to their opinions, and dealing with their complaints promptly.

Timing: Timing is an important component of change communication. This entails envisioning when and how they want to persuade their respective audiences about the necessity of change and the advantages of going digital (Ramlall 2003).

Medium: Communication during change management involves identifying the right medium through which the message will be delivered to the intended audience. This pertains to the selection of the appropriate media that is suitable for use in the communication process with the stakeholders and this may include the use of e-mail, Facebook, Twitter, or physical meetings (Sarmad et al. 2016).

Empathy: Person-to-person communication in change management has to be empathetic. This involves appreciating the emotions of others by acknowledging perceived threats and sharing them in a manner that can be easily dealt with.

6.3.2.3 Training

Training is a critical aspect of change management in supply chain management. It involves providing employees with the knowledge and skills they need to adapt to digital transformation, operate new systems and processes, and optimize supply chain operations (Hunt and Michael 1983). Here are some key aspects of training in change management:

Needs Assessment: In training in change management, one must begin by assessing the needs of the organization to determine whether the workers lack certain knowledge or skills. This involves the identification of tasks and determining the training requirements of each person based on roles, tasks, and responsibilities.

Customization: Training in change management should be customized to meet the specific needs of each employee (Arnold 2005). This involves tailoring the training program to their job role, level of expertise, and learning style.

Relevance: The change management training should have something to do with the digital transformation project. This includes increased attention to the knowledge and competencies clients require to run new systems and flows and enhance supply chain efficiency.

Hands-on Practice: Effective training in change management requires hands-on practice (Hannay and Northam 2000). This involves providing employees with opportunities to apply their new skills and knowledge in real-world scenarios and receive feedback on their performance.

Feedback and Evaluation: Change management training should also include feedback and evaluation. This includes affirming the employees giving them feedback on how they have performed in the organization, and assessing the training program that is offered to the workers.

Continuous Improvement: Effective training in change management requires continuous improvement (Kraiger 2014). This involves analyzing feedback and evaluation data and updating the training program to ensure that it remains relevant and effective.

6.3.2.4 Implementation

One of the most strategic and essential steps in change management in supply chain management is implementation. It pertains to the process of implementing the digital transformation strategy, which may refer to the implementation of new systems and tools, educating the employees about the change, and evaluation of results. Here are some important features of implementation throughout change management:

Project Management: There is no doubt that good project management is an important feature needed when implementing change management. This involves developing a project schedule based on several key activities, deciding on time frames, and defining the various roles of stakeholders.

Deployment: Implementation in change management involves deploying new systems and processes (Hersey 1996). This involves ensuring that the systems are properly installed, configured, and tested and that the processes are properly documented and communicated to employees.

Training: Training is a key factor in change management, where employees are taught how to use the new system and change. This includes tailoring the training program to ensure that it suits the capacity of the given employee, as well as ensuring that they are availed enough practice time.

Change Control: Implementation in change management requires change control (Sonnentag et al. 2004). This involves managing changes to the project plan, and ensuring that any changes are properly documented, communicated, and approved by stakeholders.

Monitoring: The concept of performance monitoring is another essential factor in implementing change management. This entails the monitoring of performance benchmarks on how the digital change initiative is performing in terms of the set objectives and goals, and finding out the gaps to be addressed.

6.3.2.5 Evaluation

In supply chain management particularly in any change process, evaluation is a vital element. It entails evaluating what has been achieved in the process of implementing the digital transformation plan and adopting improvement strategies. Here are some key considerations to the process of evaluation in change management:

Performance Metrics: In change management, performance measurements are a necessity when it comes to the evaluation process. This includes setting up key performance indicators (KPI) to measure the success of the digital transformation strategy whereby the company can monitor things like supply chain performance lead time, inventory accuracy, and customer satisfaction (Mort Feldmann et al. 2013).

Data Analysis: Whereas, assessment in change management is an analytical process. This entails reviewing and scrutinizing the data obtained from the performance metrics to recognize patterns, trends, and opportunities for improvements.

Feedback and Surveys: Evaluation in change management involves gathering feedback from stakeholders (McCauley 2001). This involves conducting surveys and collecting feedback from employees, customers, suppliers, and other stakeholders to assess their satisfaction with the digital transformation plan.

Benchmarking: Evaluation in change management involves benchmarking against industry standards (Loan-Clarke et al. 1999). This involves comparing the supply chain operations to those of competitors and other industry leaders to identify areas where improvements can be made.

Reporting: In change management, evaluation entails communicating the outcome or findings of the change process to people. This entails explaining the outcomes of the evaluation to the interested parties, which may comprise the organization's top officials, employees, clients, suppliers, and other interested parties.

6.4 Leadership and Smart SC Adoption

We share few of the examples of organizations that have successfully implemented smart SCM practices through strong leadership:

Amazon: Amazon is a prime example of a company that has leveraged technology to optimize its supply chain operations. Being one of the most famous grocers, the corporation has applied data analytics, automation and advanced robotics technologies and, therefore, has showcased great effectiveness and speed in supply chains (Modau et al. 2018). The outstanding leader of this change has been Amazon's CEO Jeff Bezos who has always encouraged the use of innovation and experimentation in the way Amazon handles its SCM.

Unilever: Unilever is one of the leading multinational consumer goods company, who have successfully integrated sustainability into their SCM strategy. Unilever has outlined its strategic direction by committing itself through Sustainable Living Plan, it's long-term visions aimed at reducing its environmental footprint, enhancing the well-being of workers and farmers in the supply chain as well as creating value for different stakeholders. There is evidence that the leadership of Unilever has been instrumental in the promotion of this sustainable SCM approach and has sought to increase the sustainability awareness of the company.

Walmart: Walmart is a retail titan that can be credited with the incorporation of technology into the SCM process and has been at the forefront of supply chain transformation for many years. Walmart's SCM practices have a root in its founder, Sam Walton who always centralized on issues to do with efficiency, cost reduction, and innovation (Das and Baruah 2013). Nowadays, Walmart remains dedicated to achieving its supply chain objectives by actively adopting different technological advancements and innovations in supply chain automation, robotics, and data analysis to enhance its effectiveness and prodigality.

Zara: Zara is a fashion retail company that operates under the fast fashion business model that is characterized by operational sensitivity to rapid changes in consumer fashion preferences. As mentioned earlier, SCM management in Zara is highly concerned with speed, agility, and responsiveness in its operation, and this segment is highly powered by the leadership team (Hassan 2022). Amancio Ortega, the CEO of the company: the company has over the years focused its supply chain strategy under the leadership of Ortega, on speed as well as flexibility in delivering the needs of customers.

6.5 Challenges to Leadership in Smart Supply Chain Management

Leadership in SCM faces several challenges that can hinder the successful implementation of smart supply chain management practices. Some of the key challenges include:

6.5.1 Resistance to Change

Effective application of smart SCM models can lead to profound changes in the company's structure and processes as well as in the use of information technologies. This can draw some opposition from employees, who may be either bureaucratic or have insufficient appreciation for the changes that are being undergone (Sandhya and Kumar 2011). These challenges therefore call for organizational leaders to enhance good communication and teamwork with their personnel. Making the employees understand the realities of the change and what gains they are likely to make out of it can assist in managing their resilient posture.

6.5.2 Lack of Data Transparency

Smart SCM focuses on leveraging data and analytical tools to drive supply chain and management decisions. Nonetheless, the fragmentation of the data sources and the inability to achieve data integration remain significant challenges for many organizations, hindering them from achieving data transparency. It can also result in the leaders lacking the proper information that will help them to manage the supply chain as efficiently as possible (Gerhart and Bretz 1994). To address this challenge, leaders of supply chain problems should ensure that data integration and visibility is enhanced throughout the supply chain. This can be through the use of data management solutions that allow for data sharing and data coordination in real-time.

6.5.3 Limited Resources

The strategies of smart SCM may entail considerable capital investments in technology, infrastructure, and human capital. However, many organizations may not have an ample amount of cash that can be invested in such activities (Gunu et al. 2013). To address this issue, leaders should focus on how to identify potential resources necessary for smart SCM programs. It may also include outsourcing some of the project aspects to partners or suppliers to tap into their resources and knowledge base.

6.5.4 Supply Chain Complexity

Supply chains in organizations are often intricate as they extend over borders, jurisdictions, and sectors. It can make it difficult for leaders across the supply chain to manage their supply chain efficiently and effectively, as well as conform to regulations and ethical standards (Nwokocha and Iheriohanma 2012). To overcome this challenge, leaders should ensure that they engage in cooperation and negotiate with their suppliers and other players. When different organizations collaborate, they get to enhance understandability through increased transparency and the reduction of intricacies in the supply chain.

To address these concerns, leaders need to focus on communication and cooperation, integration of information, distribution, and use of resources, and the development of system and partnerships (Falola et al. 2014). These challenges may, however, be overcome through strategic and collaborative supply chain planning by various organizations to foster the formation of efficient supply chains that enhance their performance within their various industries. Thus it can be seen that SSCM is emerging evidently as a need for strategic importance in the current business environments particularly to enhance the OPF and OFM. Technological and information resources, or the SSCM, is a paradigm shift that deviates from conventional supply chain management. This work further establishes the centrality of effective leadership in the implementation of SSCM strategies for practical realization. The scope of this paper is to analyze the relationship between leadership and SSCM so that the role of leaders in improving the effectiveness, integration, and innovativeness of supply chains can be elucidated. The chapter also emphasizes the leadership's importance for the implementation and development of SSCM activities. Managers hold the key to shaping a corporate culture that is open to innovation and encourages the adoption of various forms of data utilization in decision-making processes. A self-organizing work environment can enable leaders to unleash the power of teams and harness it in a proper way for supply chain optimization. The paper suggests that leadership remains the key driver of SSCM projects because the leadership can align the projects with organizational goals and priorities, and can properly authorize sustainable resources.

6.6 Conclusion

In this chapter, we have explored the importance of leadership in smart supply chain management. We have discussed how strong leadership can drive innovation, collaboration, and continuous improvement in SCM. We have also examined the benefits of smart SCM practices, including improved efficiency, reduced costs, increased sustainability, and enhanced customer satisfaction. Furthermore, we have analyzed real-world examples of organizations that have successfully implemented smart SCM practices through strong leadership, such as Amazon and Walmart. These organizations have demonstrated the positive impact that leadership can have on SCM performance, including increased efficiency, transparency, and collaboration.

However, some issues have been discussed on how leaders have faced challenges in the implementation of SCM such as: resistance to change, lack of data transparency, limited resources, and complexity of the supply chain. To effectively address the aforementioned challenges, leaders and organizations should focus on information sharing, cooperation, data linkage, resource commitment, and partnerships. The conclusion of the research emphasizes on the role of leadership in smart and its implications for organizational performance. In the modern world, which can be referred to as the era of rapid and intense competition, effective supply chain management has emerged as one of the key sources of competitive advantage. However, the viable application of smart concepts in the procurement process calls for effective leadership that is strategic, team-oriented, and vision-oriented. Leaders are instrumental in advocating for change, employee engagement, as well as the enhancement of improvement initiatives in SCM. Organizational leadership can shape the organizational culture and foster the experimentation necessary to achieve supply chain advancements. There is also the role of enhancing the visibility as well as the transparency of the chain through collaboration and partnership between different leaders for a streamlined supply of the chain. Leaders must therefore approach the management of supply chains in a proactive and integrating way, leading to the development of sustainable and optimal supply chains that will adequately address the needs of the customers and other stakeholders. The significance of leadership in determining SCM outcomes can be seen through organizations like: Amazon and Walmart to show how corporate leadership may enhance supply chain performance indicators. It has been realized that the implementation of smart SCM within these organizations has resulted in, among other benefits, reduced costs, customer satisfaction, and enhanced profitability.

Conclusively, leadership is essential for the successful implementation of smart SCM practices and has a significant impact on organizational performance. Strong leaders must prioritize communication, collaboration, data integration, resource allocation, and partnership to overcome challenges and embrace the opportunities presented by smart SCM practices. By doing so, organizations can build resilient, sustainable, and optimized supply chains that can drive success in their respective industries.

References

Arnold, E. (2005). Managing human resources to improve employee retention. *The Health Care Manager*, *24*(2), 132–140.

Asfaw, A. M., Argaw, M. D., & Bayissa, L. (2015). The impact of training and development on employee performance and effectiveness: A case study of District Five Administration Office, Bole Sub-City, Addis Ababa, Ethiopia. *Journal of Human Resource and Sustainability Studies*, *3*(04), 188.

Bag, S., Gupta, S., Choi, T.-M., & Kumar, A. (2021). Roles of innovation leadership on using big data analytics to establish resilient healthcare supply chains to combat the COVID-19 pandemic: A multimethodological study. *IEEE Transactions on Engineering Management*.

Bastidas, V., Oti-Sarpong, K., Nochta, T., Wan, L., Tang, J., & Schooling, J. (2023). Leadership for responsible digital innovation in the built environment: A socio-technical review for re-establishing competencies. Journal of Urban Management, 12(1), 57–73.

Bibi, P., Pangil, F., Johari, J., & Ahmad, A. (2017). The impact of compensation and promotional opportunities on employee retention in academic institutions: The moderating role of work environment. *Journal of Economic & Management Perspectives*, *11*(1), 378–391.

Buonocore, F., Annosi, M. C., de Gennaro, D., & Riemma, F. (2024). Digital transformation and social change: Leadership strategies for responsible innovation. Journal of Engineering and Technology Management, 74, 101843.

Campbell, J. P. (1971). Personnel training and development. *Annual Review of Psychology*, *22*(1), 565–602.

Cannon-Bowers, J. A., Tannenbaum, S. I., Salas, E., & Converse, S. A. (1991). Toward an integration of training theory and technique. *Human Factors*, *33*(3), 281–292.

Carrell, S. E. (2007). The national internal labor market encounters the local labor market: Effects on employee retention. *Labour Economics*, *14*(5), 774–787.

Chatterjee, S., & Chaudhuri, R. (2021). Supply chain sustainability during turbulent environment: Examining the role of firm capabilities and government regulation. *Operations Management Research*, 1–15.

Chatterjee, S., Chaudhuri, R., Vrontis, D., & Giovando, G. (2023). Digital workplace and organization performance: Moderating role of digital leadership capability. Journal of Innovation & Knowledge, 8(1), 100334.

Cron, W. L., Marshall, G. W., Singh, J., Spiro, R. L., & Sujan, H. (2005). Salesperson selection, training, and development: Trends, implications, and research opportunities. *Journal of Personal Selling & Sales Management*, *25*(2), 123–136.

Das, B. L., & Baruah, M. (2013). Employee retention: A review of literature. *Journal of Business and Management*, *14*(2), 8–16.

Evans, K. M., & Foster, V. A. (2000). Relationships among multicultural training, moral development, and racial identity development of White counseling students. *Counseling and Values*, *45*(1), 39–48.

Falola, H. O., Osibanjo, A. O., & Ojo, I. S. (2014). Effectiveness of training and development on employees' performance and organisation competitiveness in the nigerian banking industry. *Bulletin of the Transilvania University of Braşov*, *7*(1), 161.

Frederico, G. F., Garza-Reyes, J. A., Anosike, A., & Kumar, V. (2020). Supply Chain 4.0: Concepts, maturity and research agenda. *Supply Chain Management: An International Journal*, *25*(2), 262–282.

Gainey, T. W., & Klaas, B. S. (2003). The outsourcing of training and development: Factors impacting client satisfaction. *Journal of Management*, *29*(2), 207–229.

Gerhart, B., & Bretz Jr, R. D. (1994). Employee compensation. *New York*.

Gunu, U., Oni, E., Tsado, E., & Ajayi, O. (2013). Empirical study of training and development as a tool for organizational performance: Case study of selected banks in Nigeria. *Kuwait Chapter of Arabian Journal of Business and Management Review*, *2*(10), 78–87.

Haider, M., Rasli, A., Akhtar, C. S., Yusoff, R. B. M., Malik, O. M., Aamir, A., Arif, A., Naveed, S., & Tariq, F. (2015). The impact of human resource practices on employee retention in the telecom sector. *International Journal of Economics and Financial Issues*, *5*(1), 63–69.

Hannay, M., & Northam, M. (2000). Low-cost strategies for employee retention. *Compensation & Benefits Review*, *32*(4), 65–72.

Hassan, Z. (2022). Employee retention through effective human resource management practices in Maldives: Mediation effects of compensation and rewards system. *Journal of Entrepreneurship, Management and Innovation*, *18*(2), 137–174.

Hersey, P. (1996). Revisited. *Training & Development*, *50*, 43.

Horwitz, F. M. (1999). The emergence of strategic training and development: The current state of play. *Journal of European Industrial Training*, *23*(4/5), 180–190.

Hunt, D. M., & Michael, C. (1983). Mentorship: A career training and development tool. *Academy of Management Review*, *8*(3), 475–485.

Kossivi, B., Xu, M., & Kalgora, B. (2016). Study on determining factors of employee retention. *Open Journal of Social Sciences*, *4*(05), 261.

Kraiger, K. (2014). Looking back and looking forward: Trends in training and development research. In *Human Resource Development Quarterly* (Vol. 25, Issue 4, pp. 401–408). Wiley Online Library.

Liu, W., Wei, W., Choi, T. M., & Yan, X. (2022). Impacts of leadership on corporate social responsibility management in multi-tier supply chains. European Journal of Operational Research, 299(2), 483–496.

Loan-Clarke, J., Boocock, G., Smith, A., & Whittaker, J. (1999). Investment in management training and development by small businesses. *Employee Relations*.

A. Margherita, G. Elia, G. Solazzo, L. Gatti and A. Poorterman, "Collaborative Crisis Management: A Coordination Science Framework to Enhance Stakeholder Responses to Emergencies," in IEEE Transactions on Engineering Management, vol. 72, pp. 191-201, 2025, https://doi.org/10.1109/TEM.2024.3508037.

McCauley, C. D. (2001). Leader training and development. *The Nature of Organizational Leadership: Understanding the Performance Imperatives Confronting Today's Leaders*, 347–383.

Modau, F. D., Dhanpat, N., Lugisani, P., Mabojane, R., & Phiri, M. (2018a). Exploring employee retention and intention to leave within a call centre. *SA Journal of Human Resource Management, 16*(1), 1–13.

Mondol, E. P. (2021). The Impact of Block Chain and Smart Inventory System on Supply Chain Performance at Retail Industry. *International Journal of Computations, Information and Manufacturing (IJCIM), 1*(1).

Mort Feldmann, L., Level, A. V., & Liu, S. (2013). Leadership training and development: An academic library's findings. *Library Management, 34*(1/2), 96–104.

Murtiningsih, R. S. (2020). The Impact of Compensation, Training &Development, and Organizational Culture on Job Satisfaction and employee Retention. *Indonesian Management and Accounting Research, 19*(1), 33–50.

Nwokocha, I., & Iheriohanma, E. B. J. (2012). Emerging trends in employee retention strategies in a globalizing economy: Nigeria in focus. *Asian Social Science, 8*(10), 198.

Ramlall, S. (2003). Organizational application managing employee retention as a strategy for increasing organizational competitiveness. *Applied HRM Research, 8*(2), 63–72.

Rees, W. D., & Porter, C. (2002). The use of case studies in management training and development. Part 1. *Industrial and Commercial Training, 34*(1), 5–8.

Salas, E., Tannenbaum, S. I., Kraiger, K., & Smith-Jentsch, K. A. (2012). The science of training and development in organizations: What matters in practice. *Psychological Science in the Public Interest, 13*(2), 74–101.

Sandhya, K., & Kumar, D. P. (2011). Employee retention by motivation. *Indian Journal of Science and Technology, 4*(12), 1778–1782.

Saputra, I. K. R. D., & Riana, I. G. (2021). The relationships of work environment, compensation, and organizational commitments on employee retention. *Journal of Multidisciplinary Academic, 5*(1), 20–25.

Sarmad, M., Ajmal, M. M., Shamim, M., Saleh, M., & Malik, A. (2016). Motivation and Compensation as Predictors of Employees' Retention: Evidence From Public Sector Oil and Gas Selling Organizations. *Journal of Behavioural Sciences, 26*(2).

S. Shafique, A. Hakim, I. A. Baig and M. Mehdi, "MangoTrace: An Intelligent Supply Chain Decision Support System," 2022 International Conference on IT and Industrial Technologies (ICIT), Chiniot, Pakistan, 2022, pp. 01-04, https://doi.org/10.1109/ICIT56493.2022.9989077.

Shafique, M. N., Rashid, A., Yeo, S. F., & Adeel, U. (2023). Transforming supply chains: Powering circular economy with analytics, integration and flexibility using dual theory and deep learning with PLS-SEM-ANN analysis. Sustainability, 15(15), 11979.

Siagian, H., Jade, K., & Tarigan, Z. (2020). The role of affective leadership in improving firm performance through the integrated internal system and external integration FMCG Industry. *International Journal of Data and Network Science, 4*(4), 365–372.

Sonnentag, S., Niessen, C., & Ohly, S. (2004). Learning at work: Training and development. *International Review of Industrial and Organizational Psychology, 19*, 249–290.

Terera, S. R., & Ngirande, H. (2014). The impact of rewards on job satisfaction and employee retention. *Mediterranean Journal of Social Sciences, 5*(1), 481.

Chapter 7
Role of Knowledge Management Capabilities in Smart Supply Chains

Nosheen Rafi, Masood Nawaz Kalyar, Shamim Akhtar, and Muhammd Hamza Khan

7.1 Introduction

Supply chains have consistently faced various challenges, necessitating periodic upgrades to maintain efficiency and competitiveness. Throughout different eras, supply chains have had to adapt to changing environments, requiring innovative solutions to ensure the seamless movement of goods, information, and funds. Historically, supply chains have evolved to address these challenges, fostering collaboration and communication among partners. The challenges faced in the present era are unprecedented and multifaceted, driven by factors such as globalization, rapid technological advancements, and heightened consumer expectations. These unique pressures demand a radical transformation of traditional supply chains. The present era's challenges require the mobilization and deployment of intangible organizational resources. Knowledge management, absorptive capacity, and intellectual capital play critical roles in strengthening the adoption and implementation of smart supply chain management (SSCM). These elements ensure that organizations are equipped with the necessary skills and information to leverage technological advancements effectively.

Digital technologies have emerged as pivotal opportunities in this context. Technologies like AI, IoT, and Blockchain offer new avenues for intelligent decision-making, real-time visibility, and enhanced efficiency. Historically, technological advancements have played crucial roles in improving supply chain resilience and

N. Rafi (✉) · S. Akhtar
Department of Business Administration, Faculty of Management and Administrative Sciences, University of Sialkot, Sialkot, Pakistan
e-mail: dr.nosheenrafi@gmail.com

M. N. Kalyar
Lyallpur Business School, Government College University Faisalabad, Faisalabad, Pakistan

M. H. Khan
Hailey College of Commerce, University of the Punjab, Lahore, Pakistan

M. S. Mubarik and S. A. Khan (eds.), *Smart Supply Chain Management*,
https://doi.org/10.1007/978-981-96-1333-5_7

responsiveness. For instance, in the 1990s and 2000s, the introduction of barcode scanners for inventory tracking and Electronic Data Interchange (EDI) for information sharing marked significant milestones in automating supply chain processes (Khayyam et al. 2022). These innovations laid the groundwork for the smart supply chain, which integrates emerging technologies at every stage. In the present era, AI stands out as a transformative technology with the potential to revolutionize supply chains. AI encompasses machine learning, natural language processing, deep learning, and computer vision, optimizing each process and step within the supply chain (Holzinger et al. 2023). By offering human-level intelligence, AI is reshaping various industries, including supply chain management (Na et al. 2022). AI aids in traffic management, autonomous vehicles, product recommendations, customer service, fraud detection, automated production, quality control, and inventory management. Its ability to manage vast amounts of data, identify patterns, and make predictions enables businesses to automate their supply chain processes, from procurement to logistics.

Despite its potential, AI adoption in supply chains faces several challenges. Integrating AI with traditional supply chain processes is complex, requiring significant investments and skilled personnel (Kehayov et al. 2022; Naz et al. 2022). Additionally, data security and ethical considerations pose significant hurdles. For instance, AI-driven automation can lead to job displacement, impacting economies (Su et al. 2022). Moreover, data sharing reluctance among supply chain members can hinder the creation of a fully integrated SSC. Nevertheless, some companies have successfully harnessed AI to enhance their supply chains. Case studies highlight instances where AI improved sales configuration, production planning, quality inspection, and risk management (Helo and Hao 2022; Shah et al. 2022; Ganesh and Kalpana 2022). These examples demonstrate AI's capability to provide competitive advantages, increase profit margins, and reduce risks. However, AI adoption remains relatively low, with many applications still in their infancy (Baryannis et al. 2019).

This chapter aims to contribute to the body of knowledge by exploring the role of AI in SSCs. It addresses how AI can improve supply chain resilience, sustainability, and responsiveness, aligning with the book's overall goals. Furthermore, the chapter highlights the enablers and barriers to AI adoption in SSCs, providing insights for industry experts to enhance adoption rates and dispel misconceptions. The chapter also outlines future developments and trends for SSCs. The structure of the chapter is as follows: Sect. 7.2 discusses the motivation for this study. Section 7.3 provides an overview of SSCs in context with knowledge management. Section 7.4 presents discussions and implications. Section 7.5 concludes the chapter.

7.2 Smart Supply Chain Management as Global Protagonist

Smart supply chain knowledge management enables individuals across the supply chain to gain insights into the location of information, assess the efficiency of information flow between critical components, and identify knowledge gaps. By leveraging knowledge maps, these gaps can be addressed, ensuring seamless knowledge transfer among relevant stakeholders and driving process enhancements. Moreover, the smart supply chain system has the potential to self-organize and optimize, while real-time data from interconnected sensors throughout the facility can be integrated with user preferences for a comprehensive understanding (Demir and Paksoy 2023). By leveraging a smart system, it becomes possible to anticipate operational bottlenecks triggered by specific events. This capability enables streamlining of the system, reducing production waste and enhancing overall operational efficiency. As a result, production processes and business operations can be optimized, leading to improved efficiency and resource utilization.

In the smart supply chain, role-related experience is how people learn how to do things. Utilizing firsthand information is the most effective approach to enhancing the supply chain. Without strong knowledge management, it can be hard to document and share information so that the organization gets the most out of it. Most manufacturers keep high-risk replacement parts on hand. This often leads to higher costs and wastes valuable manufacturing space (Hwang and Seruga 2011). Smart prescriptive analytics can greatly contribute to monitoring the condition of operational equipment within a facility. In the event of malfunctioning equipment, this technology enables prompt notification to plant supervisors and facilitates expedited ordering of replacement parts. Incorporating the smart supply chain into this process further enhances its effectiveness, presenting a straightforward approach to optimizing its functionality (Kuo et al. 2021). If technology can find and fix problems with how a system works, it could make the whole system last longer.

Permitting client needs and actions Smart supply chain allows access to, sharing of, and use of data about consumer needs and behaviors, quality feedback, sales, and trends or changes in the business environment to help make decisions across different parts of the supply chain. As market demands and customer expectations rise, smart supply transparency has become a major problem for businesses (Alzoubi 2018). It is valuable to remain transparent in the supply chain when it comes to catering to environmentally aware customers, and IoT is a crucial component of this. To be specific, any disruptive incidents that may threaten the supply chain and limit the flow of goods and services can be prevented with IoT technologies. This level of transparency and IoT-based monitoring assist in building up a sustainable and robust supply chain solution.

It can also be assumed that when all sorts of information are disclosed, the efficacy and quality management of the supply chain will be enhanced. This knowledge comes from both internal and external sources, such as interactions with customers,

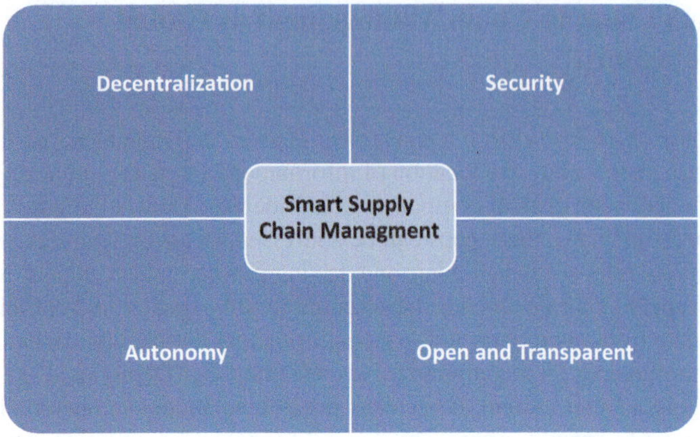

Fig. 7.1 Key paradigm of smart supply chain management. *Source* Authors own elaboration from literature

consumer and financial data, and studies of market trends and performance. Leveraging and utilizing this information can facilitate the enhancement and reevaluation of supply chain processes, leading to improvements in customer satisfaction, revenue generation, innovation, and overall product or service quality. By effectively harnessing the insights provided by this information, organizations can optimize their supply chain operations to drive positive outcomes across various dimensions, ultimately benefiting their customers and business performance (Nakasumi 2017). Knowledge management also encourages continuous learning, sharing of information, and development through collaboration. This helps your supply chain adapt to changing customer needs. Most warehouses and supply chains have written maintenance schedules that help check and fix equipment so it doesn't break down when it's needed. Even though this is the most common way to keep something in good shape, it isn't always the best. IoT makes preventive maintenance unnecessary because it lets businesses know when a piece of equipment will need to be fixed (Gupta et al. 2019). IoT sensors can help keep track of how each piece of equipment is doing and send real-time information back to the warehouse management software, which can then alert supervisors and maintenance teams. Then, the right steps can be taken to avoid expensive downtime and broken equipment. Smart supply chain management includes the following key paradigms (Fig. 7.1).

7.2.1 Decentralization

The key concept of smart supply chain technology is decentralization. Smart supply chain technology is no longer reliant on centralized processing nodes as a result of this advancement. Each node serves as a data storage, transmitter, and consumer (Benčić et al. 2019). Information storage, administration, and modification are all decentralized. When used to manage digital assets, the blockchain as a whole eliminates hierarchies and enables peer-to-peer information sharing because each node in the blockchain is subject to the same rules (Müller et al. 2019).

7.2.2 Security

In smart supply chain systems, the idea of decentralization is very important. Because of this invention, smart supply chain technology no longer needs processing hubs in the middle. Each node is in charge of storing, sending, and using information. Information storage, management, and changes are all done without a central location (Turjo et al. 2021). Because each node in the blockchain has to follow the same rules, the blockchain as a whole gets rid of hierarchies and lets people share information directly with each other when it is used to manage digital assets.

7.2.3 Autonomy

The smart supply chain may operate in an untrustworthy environment by allowing all system nodes to share, record, and update data at any time. This is primarily due to the smart chain's lack of human interaction, which shifts trust away from the person and toward the system as a whole (Calatayud et al. 2018). The fact that the account owner has complete control over his account indicates the smart chain's decentralized character. Using their private key, the user manages their account under the block's unified smart contract. Employees can now give their advice anonymously and on an equal basis.

7.2.4 Open and Transparent

The decentralized and public nature of smart technology is a result of the P2P network technology, Merkle tree technology, and consensus algorithm described above (Shah et al. 2019). Data synchronization is guaranteed by the consensus algorithm, which employs either a consensus mechanism or a reward mechanism, and peer-to-peer (P2P) network technology, which eliminates the possibility of deceit through the use of a tree-like operating structure formed by a hash algorithm.

7.3 Knowledge Management Capabilities and Smart Supply Chain

Knowledge management capabilities significantly affect smart supply chains because make it easier to share information between people and companies in a supply chain, enhance the forecast and decision of the supply chain, and supply chain productivity (Schniederjans et al. 2020). For instance, through a collection of accurate and timely information, supply chain partners can better maneuver their inventory systems, and scheduling of production and transportation network. Furthermore, knowledge management systems can be used to assess and manage risks, for example, disruptions in distribution channels, and provide opportunities for partners to share experiences, lessons learned, and new ideas. In total, the development of knowledge management capabilities can greatly assist in making the supply chain more effective and able to deliver greater added value to the customers by adapting to changing conditions.

7.3.1 Knowledge Infrastructure Capabilities and Smart Supply Chain

Knowledge infrastructure is described as the structures, tools, and procedures applied to acquire, preserve, and disseminate knowledge within an organization. Conversely, supply chain management is the process of planning, controlling, and executing activities related to the acquisition of resources for the production and distribution of products or services. The link between knowledge infrastructure and supply chain management involves the capacity that knowledge infrastructure depends on and can complement supply chain management.

Knowledge infrastructure in the decision-making context allows gaining access to the most important and relevant information within the SC environment. Supply chain management makes it easier to engage the supply chain partners and deals with the issue of lack of coordination. For example, using a knowledge management system to store and reuse necessary information including supplier data, inventory data, and delivery data can help a company manage its supply chain efficiently.

7.3.1.1 Knowledge Structure and Smart Supply Chain Management

The information that the business organization has and its information structure can greatly affect the supply chain management of the firm. The internal structures for accumulating and organizing knowledge highly influence the company's performance and its ability to respond to various changes in the environment. On the other hand, there is a possibility of coming up with a dysfunctional knowledge system that ends up complicating matters, slowing the supply chain processes down, and causing unnecessary bottlenecks. Here are some concrete accounts of how a company's knowledge structure affects SCM:

Improved communication and collaboration: Effective knowledge management can lead to the establishment of a seamless flow of insights across various teams and departments, thereby ensuring that all parties are moving in the right direction and sharing information periodically.

Better decision-making: The greatest benefit of a usable and robust search system is that managers and other decision-makers are enabled to easily find all the necessary information for their decisions.

Reduced errors and delays: Proper organization of knowledge in any organization also assists in minimizing errors and /or time consumption by providing updated and necessary information and also acknowledging any problems or changes in the supply chain.

Increased agility: When the company acquires knowledge structure, it can allow it to change much more quickly to the changes occurring in the market and meet new opportunities or challenges.

7.3.1.2 Knowledge Culture and Smart Supply Chain Management

The knowledge culture can play an impact role in improving SCM by encouraging effective communication, cooperation, and especially the commitment to improve the organization. An effective system of sharing knowledge and ideas within the workplace empowers employees in the process of finding solutions to problems as they arise. Therefore, this helps in shaping the supply chain strategies by improving efficiency and thus reducing cost thus improving customer service and competition. Secondly, a culture of learning and development guarantees that the human capital is equipped with the proper skills and knowledge to address ever-changing market conditions and remain ahead in the highly charged business environment.

Communication and Collaboration: As a result of knowledge motivation, employees are encouraged to engage in knowledge transactions, which leads to improvement in interaction and cooperation between different departments and units of the organization. This environment enables efficient identification and resolution of problems and contributes positively to the performance of the supply chain.

Continuous Improvement: Knowledge culture also promotes a climate of process improvement, where employees are encouraged to seek new ways to improve processes and systems. This commitment to continuous improvements leads to better ways of managing supply chains, better costs, improved customer satisfaction, and better competitiveness.

Adaptability: Investing in employees through continued training fosters a culture of acquisition of new knowledge to equip the employees to counter new market challenges and or compete more effectively. This is all the more true in the context of the supply chain management profession where innovations are frequent and new technologies and trends are rapidly implemented.

Innovation: A knowledge culture will also result in innovation in Supply Chain management. It means that those workers who are willing to share their ideas and knowledge can find the best ways for supply chain problem-solving and supply chain competitiveness increase.

Risk Management: With improved communication and cooperation, employees are also in a position to detect and eradicate potential risk factors that may hamper the supply chain; this may be due to factors such as natural disasters, suppliers' default, or bankruptcy.

Consequently, this study found that knowledge culture can support ECM in supply chain management by enhancing communication, collaboration, continuous improvement, adaptability, and innovation, as well as risk management. This can make the various operations involved in the supply chain easier and more efficient which in turn can be financially beneficial, lead to better customer satisfaction, and ultimately make organizations more competitive.

7.3.1.3 Knowledge Technology and Smart Supply Chain Management

Knowledge technology, such as artificial intelligence, big data analytics, and the Internet of Things (IoT), can have a significant impact on supply chain management (Fig. 7.2). Here are a few examples of how these technologies can be used in supply chain management:

Artificial Intelligence: AI could be employed to extract and make sense of data from different sources, for instance, sensors of IoT devices to enhance the efficiency of the supply chain. For instance, AI can be employed to forecast demand for products and set the right production schedule or determine the best routes for transport vehicles, hence cutting costs and emissions.

Big Data Analytics: Business intelligence solutions can be applied to process the big data from multiple sources including social networking sites for understanding the customer needs and the demand. This can be used to enhance the accuracy of the sales forecasts and inventory level and to discover more avenues of operation.

Internet of Things (IoT): IoT sensors may help to track the location and condition of items in the supply chain to provide valuable data to organizations. This can assist with the need for better tracking of products and having a way to find and solve problems more efficiently.

Blockchain: Technology known as blockchain can be employed to ensure record keeping and tracking in the supply chain strategy is enhanced and secure. It can increase the transparency of the companies' supply chain enhance credentials and decrease the likelihood of fraud.

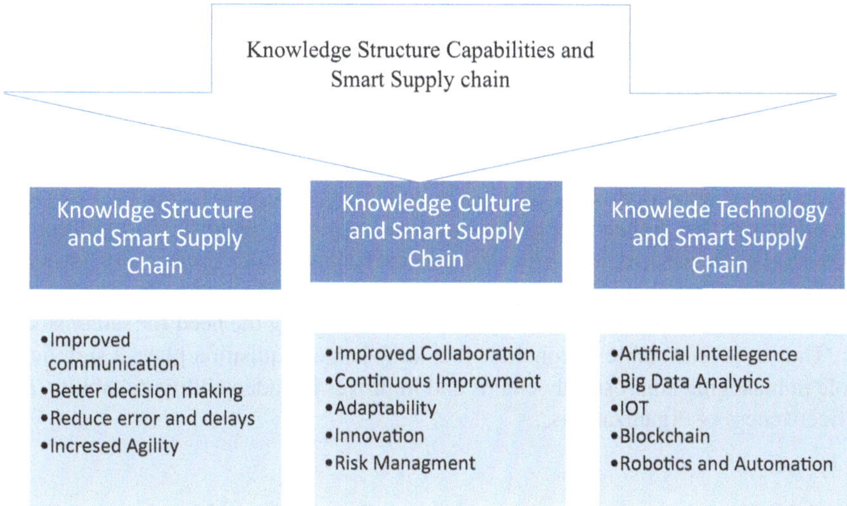

Fig. 7.2 Knowledge structure capabilities and smart supply chain. *Source* Authors own elaboration from literature

Robotics and Automation: Technologies in robotics and automation can be applied for purposes of enhancing logistics and warehousing systems. For instance, warehouse robots can help transport products and may help in the packing and picking of products and dispatching of orders.

Hence, knowledge technologies such as AI, big data, IoT, Blockchain, and automation can make a tremendous contribution to the supply chain by enhancing visibility, traceability, automation, and decision-making. These technologies can facilitate supply chain management to reduce overall supply chain costs while enhancing customer service.

7.3.2 Knowledge Process Capabilities and Smart Supply Chain

7.3.2.1 Knowledge Acquisition and Smart Supply Chain Management

One of the critical activities in smart supply chain management is the acquisition of knowledge about the supply chain and the surroundings as this will help in improving the overall performance.

Informed decision-making: This implies that organizations can plan and implement decisions effectively where they gain knowledge concerning aspects of the supply chain including supplier performance, customer demand, and market trends.

Operational efficiency: Supply chain knowledge improves the ability to recognize inefficient or blocked areas and address issues with the supply chain proactively. For instance, information on supplier lead time can assist an organization in the manner that it organizes its production cycle to ensure it avoids stock-out situations while at the same time avoiding situations whereby it has more stocks than what it can sell.

Performance improvement: It is only through learning of the SC that organizations can discover that somewhere along this line there is an opportunity to cut costs, provide better services to customers, and even make more money. For instance, stock information about customer demands can assist organizations in restructuring a production calendar to reflect demand, thus eliminating the need for safety stock.

Thus, it is reasonable to conclude that knowledge acquisition plays a significant role in managing smart supply chains and improves the adaptability, reliability, and effectiveness of organizations.

7.3.2.2 Knowledge Conversion and Smart Supply Chain Management

Knowledge conversion is also central to smart supply chain management because organizations use data and information to convert them into explicit and tacit knowledge, for creating value. In other words, knowledge conversion can affect the SSCM in the following ways:

Data Management: It enables organizations to organize data from different data sources like sensors, RFID tags, and EDI systems more effectively and efficiently by integrating converted data structures into integrated data structures.

Insights Generation: Transforming data into knowledge is crucial as it provides superior comprehension of supply chain performance and offers an understanding of where improvements need to be made.

Predictive Analytics: By translating data into knowledge, an organization can employ predictive modeling for demand forecasting, risk assessment, and inventory management.

Automation: The conversion of data to knowledge helps organizations to implement automation technologies including Robotics and Drones for effectiveness in repetitive chores.

Blockchain: When the data is turned into knowledge, through the help of blockchain technology, organizations can establish trusted networks for sharing data and synchronizing activity along the supply chain. Therefore, dynamic knowledge conversion plays a critical role in the effective implementation of smart supply chain management. It refers to the process of acquiring data and turning it into a usable form in the organization to help identify patterns, improve processes, and enhance value from the supply chain.

7.3.2.3 Knowledge Application and Smart Supply Chain Management

A smart supply chain entails the precise application of newer technologies including, IoT, AI, and blockchain within the supply chain environment for purposes of value addition. Knowledge application plays a critical role in smart supply chain management, as it enables organizations to:

Collect and Analyze Data: The organization can then use the available knowledge on techniques of collecting and analyzing collected data from sources like sensors, RFID tags, and EDI systems to get an insight into the performance of the supply chain.

Predictive Analytics: The concepts of machine learning and predictive analytics, for example, allow predicting demand, risks, and current stock levels.

Automation: By analyzing forms of automation, including robotics and drones, it is possible to understand how automation can be utilized to enhance efficiency within an organization.

Blockchain: Majoring in the knowledge of blockchain, organizations can develop a secure decentralized network, where data are transferred and activities organized throughout the supply chain.

Network Optimization: Applying optimal flow strategies makes it possible for various organizations in the chain to balance and re-direct goods and information to achieve better efficiency and save costs.

Hence knowledge application plays a crucial role in the implementation of smart supply chain management. It allows organizations to derive the most optimum value of newer technologies to develop newer value propositions, optimize performance, and abate risks.

7.3.2.4 Knowledge Protection and Smart Supply Chain Management

Knowledge protection also has a significant role to play in smart supply chain management as it assists companies in protecting their knowledge and innovation. This can in turn have a considerable influence on the efficiency and competency of the supply chain.

Maintaining competitive advantage: Knowledge protection assists in ensuring that the organization's information and intellectual capital are safeguarded to enable the organization to leverage competitive advantage within the markets. For instance, through the adoption of patents, organizations can extend legal protections over product designs and methods of production to discourage competitors from mimicking their products.

Enhancing trust and collaboration: This practice can also improve trust and cooperation within the supply chain since information is released to suitable individuals only. This can tend to enhance communication as well as coordination among the partners in effecting improved supply chain system.

Compliance with regulations: Knowledge preservation is also crucial due to various industries' legal requirements like data privacy laws. Businesses have to make sure that personal information and sensitive information are being protected and that they are adhering to all the rules and regulations in this matter.

Risk management: Through 'data masking' plans, the vulnerability of organizations to risks of leakage of private information, frustration of cyber criminals, loss of funds, and tarnished image can be minimized. Some of the benefits of protecting knowledge include: knowledge protection allows organizations to recognize and manage any risks that may be inherent in its transfer and sharing process.

Therefore, knowledge protection can be considered one of the major components of the concept of smart supply chain management because it allows organizations to protect their valuable knowledge and intellectual property. This can assist them in sustaining their competitive edge, improve the interconnection and cooperation between the partners, as well as ensure the satisfaction of legal requirements and protection against data leaks and cyber threats (Fig. 7.3).

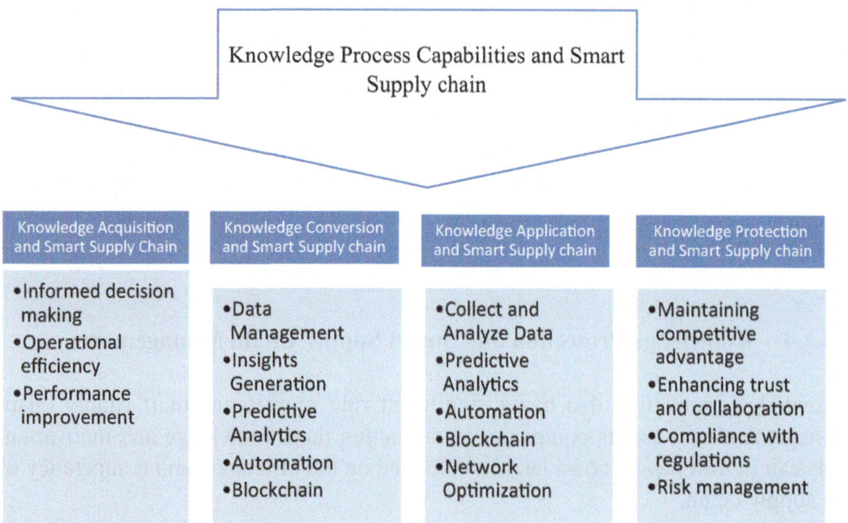

Fig. 7.3 Knowledge process capabilities and smart supply chain. *Source* Authors own elaboration from literature

7.4 Discussion

Knowledge management can assist your supply chain enhance product quality, customer service, sales, and task efficiency over time. Good knowledge management in supply chain businesses may promote knowledge exchange and innovation. Academics and business people are interested in distributed ledger technology (DLT) because of recent research, projects, and conversations. DLT could replace the trust architecture that has plagued centralized systems for decades. Traditional company operations rely on central authorities or third parties like banks to generate confidence among participants. But individuals who intended to hurt them attacked, broke, and transformed them. Decentralized blockchain technology has no central authority. This may require business reorganization. Animals hoard food to survive winter. Gray squirrels barely recover 74% of their buried food. Some companies stockpile industrial automation system spare parts in case equipment breaks down. Modern technology can reduce waste and optimize industrial space to speed up this process. Siemens and the German government modeled Industry 4.0 after the "smart factory." Smart factories use advanced analytics to reduce downtime and waste and integrate all building systems. Lean manufacturing is an old concept. Reduces industrial waste. Taiichi Ohno, a Toyota manager, created it shortly after World War I. Fear exists among manufacturers that they will fall further behind rivals who have adopted smart and cutting-edge manufacturing technologies to compete globally. The smart supply chain system will organize and optimize its operations automatically. The data gathered by factory-connected sensors can be combined with user preferences, the local climate, and any other variables permitted by the system. A smart system is capable of anticipating a bottleneck caused by an unusual weather event. When such occurrences are anticipated, the system can be streamlined and excessive production waste can be avoided, resulting in leaner production. Integration of machine learning, radio frequency identification (RFID), and consignment policy can help to reduce unreliability in smart supply chain management by providing real-time visibility into inventory levels, tracking the movement of goods, and automating decision-making processes. Machine learning algorithms can be used to analyze data from RFID systems and predict demand patterns, helping to optimize inventory levels and reduce stockouts. RFID technology can be used to track the movement of goods, providing real-time visibility into inventory levels and helping to identify bottlenecks in the supply chain. A consignment policy, which involves holding inventory at a customer site rather than at a central warehouse, can help to reduce the risk of stockouts and improve responsiveness to changes in demand. By integrating these technologies and policies, companies can improve supply chain efficiency, reduce costs, and enhance customer satisfaction.

At the retail giant Gap, inventory management focuses on having the right products in the right quantities at all times. This demonstrates the incorporation of a smart supply chain into a business model. The company can monitor inventory and track goods down to each item, thereby enhancing the user experience and reducing costs. The operating environment of the company is extremely competitive.

Typically, manufacturers will keep a supply of replacement components on hand, especially if those components are deemed the most likely to fail. Even though it occupies valuable floor space in the factory, a vast quantity of unused equipment is permitted to gather dust and sit idle. Management may be concerned that replacement components for defective machinery will be difficult to get due to the acceleration of industrial obsolescence. This is especially important for businesses that have outdated systems. The condition of every plant operating equipment can be monitored using predictive analytics. If equipment is not performing optimally, plant managers may receive automatic notifications. Before the entire technique fails, a new part is frequently fitted. Incorporating the smart supply chain into this process is a simple way to achieve even greater improvements. If technology can detect and monitor a problem in a system and then make the appropriate adjustments, the system's overall lifespan could be extended. When it is not possible to update the system, a smart factory may order a new component from a reliable source of used industrial parts. Manufacturers would no longer need to keep excess parts on-site, reducing waste and increasing industrial space.

This type of smart supply chain management eliminates the need to take seasonal availability into account. Throughout the year, you will have access to replacement parts. We are all aware that winter is on its way, but producers who use the smartest methods, procedures, and technology will make it through the spring without producing as much trash as the gray squirrel with its unrecovered resources.

7.5 Conclusion

Knowledge management and smart supply chain management can work together to improve a business by utilizing data and information to optimize processes and make better decisions. By implementing knowledge management techniques, a company can gather, organize, and share relevant information and expertise within the organization. This can help to improve communication and collaboration among employees and can aid in decision-making, problem-solving, and innovation.

When combined with smart supply chain management, this knowledge can be used to improve the efficiency and effectiveness of the supply chain. For example, by using data and analytics, a company can optimize inventory levels, improve forecasting, and make more informed decisions about suppliers and logistics. Additionally, by sharing knowledge among partners in the supply chain, such as suppliers and logistics providers, a company can improve the coordination and collaboration of the entire supply chain, leading to further efficiencies and cost savings.

7.5.1 Managerial Implications

Managers can apply smart supply chain management (SCM) in conjunction with knowledge management to improve the efficiency and effectiveness of their supply chain operations. Smart supply chain management with knowledge management can improve the efficiency and effectiveness of their supply chain operations by leveraging data and analytics, improving communication and coordination, documenting and sharing best practices, and transferring expertise within the organization. Managers can apply smart SCM along with knowledge management by leveraging data and analytics tools to gain insights into their supply chain operations. By collecting and analyzing data on suppliers, inventory levels, and customer demand, managers can gain a better understanding of their supply chain and make more informed decisions. Another way managers can practice smart SCM with knowledge management is by implementing collaborative platforms to share information and best practices with suppliers and partners. This can help to improve communication and coordination across the supply chain, which can lead to more efficient and effective operations. Managers can also practice knowledge management tools to document and share best practices and lessons learned within the organization. It helps to improve the efficiency of supply chain operations by identifying and replicating successful processes. In addition, managers can use knowledge management to identify and transfer the expertise of key employees, particularly those who are retiring or leaving the company, to ensure continuity of operations.

References

Alzoubi, H. M. (2018). The role of intelligent information system in e-supply chain management performance. *Intelligent Information System Supply Chain, 7*(2), 363–370.

Baryannis, G., Validi, S., Dani, S., & Antoniou, G. (2019). Supply chain risk management and artificial intelligence: State of the art and future research directions. *International Journal of Production Research, 57*(7), 2179–2202.

Benčić, F. M., Skočir, P., & Žarko, I. P. (2019). DL-Tags: DLT and smart tags for decentralized, privacy-preserving, and verifiable supply chain management. *IEEE access, 7*, 46198–46209.

Calatayud, A., Mangan, J., & Christopher, M. (2018). The self-thinking supply chain. *Supply Chain Management: An International Journal, 24*(1), 22–38.

Demir, S., & Paksoy, T. (2023). Fundamental Concepts of Smart and Sustainable Operations and Supply Chain Management. In *Smart and Sustainable Operations and Supply Chain Management in Industry 4.0* (pp. 1–26): CRC Press.

Ganesh, A. D., & Kalpana, P. (2022). Future of artificial intelligence and its influence on supply chain risk management–A systematic review. *Computers & Industrial Engineering, 169*, 108206.

Gupta, S., Drave, V. A., Bag, S., & Luo, Z. (2019). Leveraging smart supply chain and information system agility for supply chain flexibility. *Information Systems Frontiers, 21*, 547–564.

Helo, P., & Hao, Y. (2022). Artificial intelligence in operations management and supply chain management: An exploratory case study. *Production Planning & Control, 33*(16), 1573–1590.

Holzinger, A., Saranti, A., Hauschild, A. C., Beinecke, J., Heider, D., Roettger, R., ... & Pfeifer, B. (2023, August). Human-in-the-loop integration with domain-knowledge graphs for explainable federated deep learning. In *International Cross-Domain Conference for Machine Learning and Knowledge Extraction* (pp. 45–64). Cham: Springer Nature Switzerland.

Hwang, H. J., & Seruga, J. (2011). An intelligent supply chain management system to enhance collaboration in textile industry. *International Journal of u-and e-Service, Science and Technology*, 47–61.

Kehayov, M., Holder, L., & Koch, V. (2022). Application of artificial intelligence technology in the manufacturing process and purchasing and supply management. *Procedia Computer Science, 200*, 1209–1217.

Khayyam, S., Ramish, A., Rehman, K., & Syed, A. (2022). Strategizing radio frequency identification (RFID) in the retail supply chains of pakistan: A multiple case study. *Operations and Supply Chain Management: An International Journal*, 15(1), 27-40.

Kuo, T.-C., Chen, K. J., Shiang, W.-J., Huang, P. B., Otieno, W., & Chiu, M.-C. (2021). A collaborative data-driven analytics of material resource management in smart supply chain by using a hybrid Industry 3.5 strategy. *Resources, Conservation and Recycling, 164*, 105160.

Li, N., Yan, Y., Yang, Y., & Gu, A. (2022). Artificial intelligence capability and organizational creativity: The role of knowledge sharing and organizational cohesion. *Frontiers in Psychology, 13*, 845277.

Müller, M., Garzon, S. R., Westerkamp, M., & Lux, Z. A. (2019). *HIDALS: A hybrid IoT-based decentralized application for logistics and supply chain management.* Paper presented at the 2019 IEEE 10th Annual Information Technology, Electronics and Mobile Communication Conference (IEMCON).

Nakasumi, M. (2017). *Information sharing for supply chain management based on block chain technology.* Paper presented at the 2017 IEEE 19th conference on business informatics (CBI).

Naz, F., Kumar, A., Majumdar, A., & Agrawal, R. (2022). Is artificial intelligence an enabler of supply chain resiliency post COVID-19? An exploratory state-of-the-art review for future research. *Operations Management Research, 15*(1), 378–398.

Schniederjans, D. G., Curado, C., & Khalajhedayati, M. (2020). Supply chain digitisation trends: An integration of knowledge management. *International Journal of Production Economics, 220*, 107439.

Shah, B., Dutta, P., Gunasekaran, A., & Kansara, S. (2022). Guest editorial: Sourcing strategies and supply chain operations under the impact of COVID-19. *Journal of Global Operations and Strategic Sourcing, 15*(2), 145–150.

Shah, S., Logiotatopouloh, I., & Menon, S. (2019). Industry 4.0 and autonomous transportation: the impacts on supply chain management. *International Journal of Transportation Systems, 4*.

Su, E., & Daspit, J. (2022). Knowledge management in family firms: A systematic review, integrated insights and future research opportunities. *Journal of Knowledge Management, 26*(2), 291–325.

Turjo, M. D., Khan, M. M., Kaur, M., & Zaguia, A. (2021). Smart supply chain management using the blockchain and smart contract. *Scientific programming, 2021*, 1–12.

Chapter 8
Emerging Technologies and Smart Supply Chain Management: A Roadmap for Future Success

Syed Ahsan Ali Zaman, Habiba Qazi, Arooba Zafar, and Yushi Jiang

8.1 Introduction

Supply chain networks have existed for a considerable period, and over time, technological advancements have emerged in various fields. The development of Industry 4.0 technologies, such as the Internet of Things, augmented reality, and advanced robotics, has led to increased intelligence in supply chain networks (Ammar et al. 2022). The supply chain management landscape is undergoing a rapid and profound transformation due to various technological innovations, such as artificial intelligence (AI), automation, blockchain, and IoT (Centobelli et al. 2020). These advancements have the potential to revolutionize how goods are produced, transported, and delivered, creating new business opportunities to increase efficiency, productivity, and profitability (Zaman et al. 2023). This chapter explores how emerging disruptive technologies (EDTs) are transforming traditional supply chain processes, making them more intelligent, automated, and innovative. The focus is on how these technologies can be leveraged to enhance supply chain performance and ensure future success. The primary aim of this chapter is to provide a comprehensive understanding of the impact of EDTs on supply chain management, highlighting the benefits, challenges, and future trends. By examining the integration of these technologies, the chapter seeks to offer a roadmap for businesses to navigate the evolving landscape and maintain a competitive edge.

S. A. A. Zaman (✉)
School of Economics and Business, Digitalization Research Group, K. Donelaicio G. 20, 105, 44239 Kaunas, Lithuania
e-mail: syed.zaman@ktu.edu

S. A. A. Zaman · Y. Jiang
School of Economics and Management, Southwest Jiaotong University, Chengdu, China

H. Qazi · A. Zafar
Department of Business Administration, Jinnah University for Women, Karachi, Pakistan

M. S. Mubarik and S. A. Khan (eds.), *Smart Supply Chain Management*,
https://doi.org/10.1007/978-981-96-1333-5_8

Technological advancements offer a range of benefits, such as increased efficiency, greater visibility, and improved sustainability and digitalization performance (Zaman et al. 2023, 2024a, b). For example, automation and robotics can reduce the need for manual labor, lower error rates, and improve productivity (Aamer et al. 2023). AI-powered algorithms can optimize route planning and demand forecasting. Blockchain can help ensure the integrity and traceability of supply chain transactions, making it easier to identify and address issues (Hu et al. 2023). Smart contracts can automate payments and reduce fraud. IoT devices can track inventory levels and monitor equipment performance in real time, improving supply chain transparency and reliability. IoT-enabled sensors can provide real-time data on shipment conditions. Big data analytics can provide insights into consumer behavior, optimize inventory levels, and predict market trends. Augmented reality can improve warehouse operations by providing real-time information to workers and enhancing training programs.

The overview discusses how EDTs are transforming traditional supply chain processes, making them more intelligent, automated, and innovative, and how adopting a smart supply chain can improve overall supply chain performance (Rodríguez-Espíndola et al. 2020). As we delve deeper into the technological revolution, it becomes evident that modernizing outdated systems is crucial for staying competitive. Many traditional supply chain systems are inefficient and lack the integration needed to handle modern demands. Integrating AI, blockchain, IoT, big data, and augmented reality can streamline operations and improve efficiency. Case studies of companies that have successfully modernized their supply chains demonstrate the benefits of adopting advanced technologies. Looking ahead, several trends and predictions will shape the future of supply chain management. In the short term (next 3–5 years), the adoption of digital technologies will accelerate, with more companies leveraging AI and IoT for real-time decision-making. In the long term (beyond 5 years), advanced technologies like quantum computing and 5G will further revolutionize supply chain operations. While these advancements offer significant opportunities, they also present challenges that businesses must navigate to remain competitive (Wiechmann et al. 2022).

Despite the numerous benefits, integrating technology in supply chain management poses various challenges. Integrating technology within the supply chain generates substantial data, which may compromise security and privacy (Agarwal et al. 2022). Enterprises must guarantee the security of their systems and safeguard their data against unauthorized access and cyber threats. Incorporating automation, AI, and other technological advancements necessitates the acquisition of novel skill sets and expertise. Enterprises must allocate resources toward training and development programs to equip their workforce with the necessary competencies (Mosteiro-Sanchez et al. 2022). Integrating technology can give rise to regulatory and compliance concerns. Enterprises must ensure that their systems adhere to pertinent legal and regulatory frameworks (Rawat 2022). Integrating novel technologies may necessitate substantial investments in infrastructure and systems. Businesses must carefully consider the costs and benefits and ensure they have the resources and expertise to implement and manage these systems effectively (Ajayi and Laseinde 2023). While integrating technology can improve transparency and traceability, it can pose new challenges around data sharing and ownership. Businesses must develop clear policies and protocols around data sharing and ownership to manage their supply chain effectively (Fekri et al. 2022).

By addressing these key issues and challenges, businesses can successfully integrate technology into their supply chain management, improving efficiency, productivity, and profitability while managing risks and ensuring compliance (Zaman et al. 2023, 2024a, b). In conclusion, readers will gain a comprehensive understanding of technology's impact on the supply chain industry. They will gain insights into the latest trends and innovations in the field and how these technologies transform traditional supply chain management practices. People will gain a holistic understanding of the future of the supply chain in the presence of the technological revolution and the opportunities and challenges that lie ahead. They will be better equipped to make informed decisions and take actionable steps to integrate technology into their supply chain operations, improving their overall market performance and competitiveness.

In this book chapter, we will thoroughly discuss all aspects of the technological revolution, with a strong emphasis on the impact of both current and future industrial revolutions on the supply chain. Starting from the historical context and significance of the technological revolution in the supply chain, we will delve into the industrial revolution and its relevance to the present and future. Readers will gain valuable knowledge about these revolutions, their transformative effects, and their implications for the supply chain industry, providing a comprehensive understanding of the ongoing and upcoming changes in this dynamic field.

8.2 Background

The industrial revolution is relevant to the background of this book chapter. The following content will explain the revolutions of industries from time to time.

Technological advancements also influence human production methods. The transition to production technology, which marked a significant departure from previous methods, is commonly called the industrial revolution. According to (Desoutter 2022), implementing novel production technologies has significantly altered individuals' working conditions and lifestyles. The Industrial Revolution represents a significant technological transformation that has profoundly impacted human life and work processes. Integrating information technology into daily life has resulted in digital advancements that have influenced various scientific disciplines (Ellitan 2020). The industrial sector experiences perpetual transformation. The industrial revolution is categorized into four phases: Industry 1.0 to Industry 4.0 (Fig. 8.1). The industry's history has witnessed three distinct phases, while the present is currently undergoing the fourth phase, commonly referred to as Industry 4.0 (DiConneX 2020). Each instance of the industrial revolution has presented advantages and disadvantages to the socioeconomic conditions of the nations involved. The various industrial revolutions have led to economic expansion, heightened efficiency, and improved

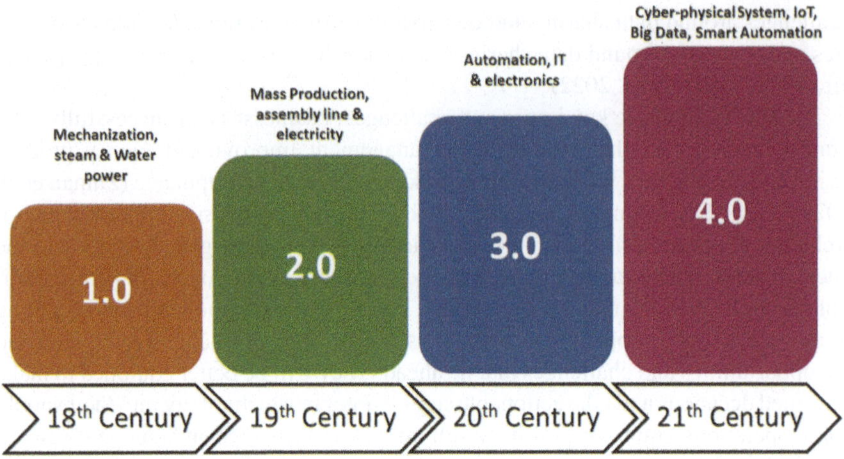

Fig. 8.1 Stages of industrial revolutions. *Source* Author

well-being in nations that have successfully harnessed their favorable outcomes, producing superior goods and services (Raja Santhi and Muthuswamy 2023).

8.2.1 *The Era of Mechanization*

The initial Industrial Revolution in 1765 marked a significant shift from an agrarian economy to one dominated by mechanization and industry. This transformation aimed to minimize reliance on human labor, replacing it with mechanized processes, facilitated mainly by the innovation of steam power (Har et al. 2022; Lee et al. 2022). The discovery of steam engines, fueled by coal, transformed manufacturing and transportation, impacting numerous sectors such as textiles, iron, farming, and mining and leading to the emergence of new industries (Amesite 2021; Geographic 2022; Özdağoğlu and Bahar 2022; Sime 2021). Steam power became a symbol of technological advancement, defining an era characterized by mass production and improved quality of life.

8.2.2 *Piece Work and Assembly Lines*

The Second Industrial Revolution, beginning in the early 1900s, was marked by the advent of electricity, telegraph, and telephone, advancing machine tools to enable mass production across various sectors, including iron, steel, transportation, chemical engineering, and telecommunications (Bühler et al. 2022; Frauendorf and de Souza 2023; Vinitha et al. 2020). Electricity replaced steam power, allowing for the implementation of assembly line labor and facilitating large-scale production at unprecedented speeds (DiConneX 2020; Lee et al. 2022). This era was characterized by enhanced efficiency in mass production and expedited transportation and communication systems, such as telegraph and railroad networks. Innovations in management and the widespread use of assembly lines further improved productivity (Muhammad et al. 2022; Simon 2023).

8.2.3 Automation Through Computers

The third industrial revolution, or Industry 3.0, began in the twentieth century, characterized by the advancement of electronics and automation through devices such as transistors, integrated circuits, and microprocessors. The focus was on efficiency, speed, and precision, leading to the development of automated machines that reduced human labor (Butt et al. 2020; Dekavadiya 2021). This era saw the rise of large-scale manufacturing using digital logic, leading to innovations such as computers, cell phones, and information technology, which significantly influenced various sectors such as manufacturing, communication, and healthcare (Melnyk et al. 2023; Tallera 2022). Techniques such as 3D printing emerged, and companies such as Microsoft and Apple were founded, reflecting the era's emphasis on automation and information technology (Jiang et al. 2022; Sahoo and Lo 2022; Zonar 2023). Industry 3.0 continues to be prevalent among micro, small, and medium enterprises (Luo and Lin 2021).

8.2.4 Digitalization and Networking

Industry 4.0, the fourth industrial revolution, has marked a significant shift in business operations by integrating big data analytics, cloud computing, augmented reality, the Internet of Things, blockchain technology, and industrial simulation (Sharma et al. 2022). These advancements have greatly influenced manufacturing, enabling enhanced flexibility, efficiency, productivity, innovation, and competitiveness (Rad et al. 2022). Additionally, Industry 4.0 may contribute to the sustainability of industrial systems (Bai et al. 2020). With an emphasis on automation, management, and instant communication through a blend of modern technologies, Industry 4.0 represents a significant trend in the integration of information technology and robotics in manufacturing (Kumar et al. 2022). The proliferation of advanced technologies integral to Industry 4.0 is on the rise, and a collection of these key technologies is depicted in Fig. 8.3.

8.3 Literature Review

The utilization of technology has numerous effects on supply chain management, resulting in increased operational efficiency for businesses. The primary advantages of technology implementation in supply chain management are cost reduction, customer service enhancement, and operational efficiency augmentation. The advent of technology in the supply chain industry has resulted in significant disruptions to the established order, leading to enhanced streamlining and simplification of processes (Aeologic 2022; Zaman et al. 2024a, b). Currently, digital business is at the forefront of widespread consideration. The term "digital" is often associated with marketing or information technology. However, it is important to note that digital technologies have the potential to transform the management and design of supply chains significantly (Ancillai et al. 2023). Supply chain management aims to ensure that the appropriate quantity of products and services are accessible to the customer at the correct cost and time. Emerging technologies are expected to significantly impact the conventional approach to delivering products and services to consumers (Park and Li 2021).

Conventional supply chains are becoming progressively intelligent by integrating sensor-based objects, enhanced connectivity, intelligent decision-making, and automation technologies. The contemporary intelligent supply chain presents compelling motivations for cost reduction and productivity enhancement. Supply chain management refers to the coordination and management of activities involved in producing and delivering goods and services, from the initial sourcing of raw materials to the final delivery of the finished product to the end customer (Andronie et al. 2021). Supply chain management involves the comprehensive management of the flow of products or services to optimize efficiency, distribution, customer satisfaction, and profitability (Rosairo 2021). The optimization of the supply chain is a critical procedure, as it leads to reduced expenses and enhanced productivity in the production cycle, as depicted in Fig. 8.2. Enterprises endeavor to enhance their supply chains to diminish expenses and sustain competitiveness (Hayes 2022). To optimize operational efficiency, an intelligent supply chain leverages a variety of nascent technologies, such as big data, the Internet of Things (IoT), blockchain, and robotic process automation (RPA). Implementing these technologies enables supply chain enterprises to curtail expenses, abbreviate delivery durations, diminish adverse ecological consequences, and attain unparalleled degrees of mechanization. A crucial aspect of an intelligent supply chain is its ability to enhance itself and demonstrate resilience in an uncertain setting. According to Davydov (2022), an intelligent supply chain necessitates the smooth exchange of information, partial dependence on automation, and ongoing enhancement of workflows by utilizing real-time data.

The concept of supply chain management (SCM) pertains to the optimal provision of goods or services, wherein the appropriate item is made available in the correct quantity, at the precise time and location, and at a reasonable cost while ensuring its quality and meeting the needs of the intended recipient (Mohammed et al. 2023).

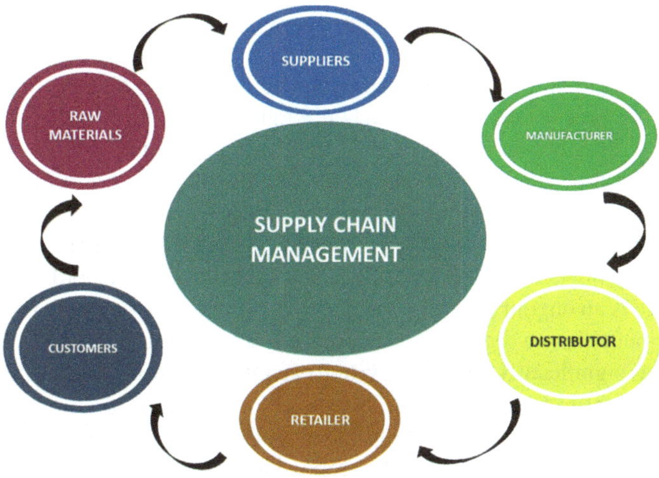

Fig. 8.2 Supply chain management process. *Source* Author

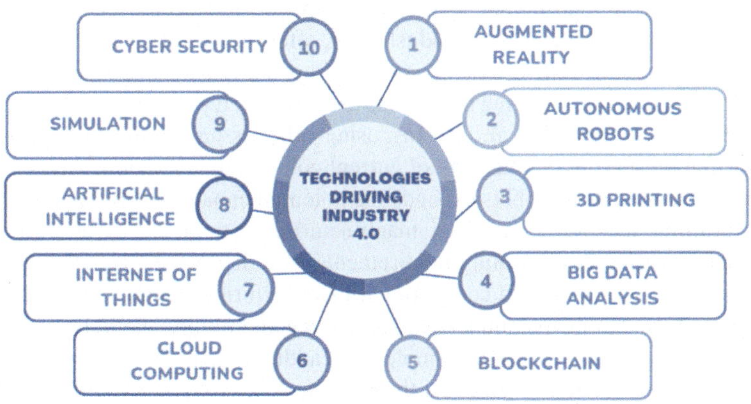

Fig. 8.3 Major technologies backing Industry 4.0. *Source* Author

Supply chains often have numerous supply–demand mismatch issues, including overstocking, stock out, and delivery delays. Due to the intricate and uncertain nature of supply chains, these problems have been extensively studied in the business and management literature. Supply chains must exhibit heightened intelligence to effectively address the growing challenges (Bhattacharyya et al. 2021). The latest iteration of the supply chain leverages advancements in semiconductors, computer science, and other engineering domains to create a comprehensive intelligent infrastructure that integrates data, information, physical objects, products, and business processes on a large scale (Yang et al. 2021). Factories equipped with advanced equipment and instruments can efficiently fulfill orders through global teams, intelligent analytics, and dynamic systems operating across various value chain stages (Zhang et al. 2023). The concept of a smart supply chain is centered on optimizing decision-making processes, leveraging data acquired through the Internet of Things (IoT) devices, and delivering precise analysis of goods throughout the entire supply chain, from production to distribution (Lazaroiu et al. 2022). Implementing smart supply chain technology would enable consumers to automate distribution and fulfillment, accurately predict the real-time status of goods, and monitor critical data that contribute to the overall efficiency of the supply network. Smart supply chains are expected to significantly impact economic growth in the coming years. According to Rosairo (2021), there is an ongoing paradigm shift in supply chain processes from linear and sequential methods to a digitized, open platform segment of Industry 4.0 (Fig. 8.3). This shift has significantly increased system capability, allowing for high throughput and efficient use of scarce resources.

The utilization of augmented reality enhances the communication of a real-world environment by incorporating virtual perceptual information, often through multiple sensory modalities, such as visual and auditory, to amplify the existing physical elements. According to Vinitha et al. (2020), augmented reality has been implemented across various industries, such as entertainment, tourism, marketing, surgery, logistics, manufacturing, and maintenance. Autonomous robots represent a significant technological advancement within Industry 4.0, offering a wide range of capabilities in manufacturing. The utilization of this technology has resulted in the augmentation of automation systems, enabling the precise execution of repetitive tasks at a reduced cost. According to Javaid et al. (2021a), using robotics has produced high-quality goods while preserving the integrity of current collaborative systems. According to (Rupp et al. 2021), using robots expands the potential capabilities of humans.

3D printing technology enables manufacturing enterprises to produce their components with reduced tooling requirements, at a decreased expense, and at an accelerated pace compared to conventional manufacturing methods. In contrast to conventional machining techniques such as drilling and milling that rely on subtractive material removal, 3D printing employs an additive manufacturing process that involves the construction of products through the successive layering of materials. This obviates the need for component and part assembly, as noted by Raji et al. (2021).

Big data technology enables the systematic extraction and analysis of information from vast and intricate datasets. The process encompasses a range of activities, such as data capture, analysis, storage, transfer, search, visualization, and updating of data queries. "Big data" refers to the aggregation of vast amounts of structured, semi-structured, and unstructured data within a singular repository. The analysis and processing of data are facilitated by integrating physical devices with the Internet, as described by (Vinitha et al. 2020). The implementation of blockchain technology has the potential to enhance security measures, safeguard privacy, and promote data transparency for both small and large businesses. The fourth industrial revolution, known as Industry 4.0, amalgamates novel manufacturing techniques that expedite manufacturers' attainment of production objectives. Blockchain technology facilitates the creation of a tamper-proof and immutable ledger shared among a consortium of business partners, enabling near-instantaneous data replication across the network. According to (Javaid et al. 2021b), the procedure involves retrieving data formerly archived in the organization's Enterprise Resource Planning (ERP) system.

Cloud computing refers to remotely storing and retrieving data, software, and services on servers via the Internet. Companies can utilize computer resources without developing the necessary infrastructure. The ability of cloud technology to provide access and processing of data without any location or organizational constraints is referred to as cloud capability. According to (nibusinessinfo 2022), Industry 4.0 leverages the cloud to provide real-time information and scalability, facilitating the integration of numerous connected devices and sensors and managing the vast amounts of data they generate.

The Industrial Internet of Things (IIoT) is a crucial component of Industry 4.0, so the two concepts are frequently utilized synonymously. In the context of Industry 4.0, many physical entities, such as devices, robots, machinery, equipment, and products, are equipped with sensors and RFID tags. These technological features enable real-time data regarding entities' condition, performance, or location. According to (Burrus 2022), this technology facilitates the optimization of supply chain operations, expedites the process of product design and modification, mitigates equipment downtime, enables companies to stay abreast of consumer preferences, and facilitates product and inventory tracking, among other benefits.

AI facilitates decision-making and detects malfunctions. Industry 4.0 integrates AI with technology to enhance efficiency in manufacturing. This burgeoning advancement improves quality and reduces operational expenses (Kar et al. 2022). The smart factory is characterized by interconnected production processes involving multiple machines capable of communicating Javaid et al. (2022). Artificial intelligence and machine learning are crucial in the paradigm shift toward Industry 4.0, facilitating the transition to intelligent factories and advanced manufacturing processes. According to (Microsoft 2022), artificial intelligence (AI) leverages the data generated by smart factories to enhance machinery performance, optimize workflows, and streamline logistics processes, ultimately leading to increased operational efficiencies.

The data generated by IoT sensors on the factory floor are now used to run detailed simulations that completely change the game's rules. Manufacturers can use technologies such as digital twins to create digital copies of their entire systems and test them in various settings and environments. Industry 4.0 offers a predictive approach to most key processes, allowing companies to optimize production and perfect existing methods. All simulations are built to be highly accurate and realistic, and detailed simulations help companies save considerable time and money (Wizata 2022).

Industry 4.0 emphasizes connectivity, so cybersecurity is certainly one of the primary concerns. Manufacturers must find ways to protect their data and systems from cybercriminals. One wrong move or questionable measure can damage the system, machinery, and brand. That is why cybersecurity is becoming more important than it was in the past (Wizata 2022).

The fourth industrial revolution—also named Industry 4.0—is one of the most trending topics in both professional and academic fields. It also considers the integration of the factory with the entire product lifecycle and supply chain activities, changing even how people work (Hernandez-de-Menendez et al. 2020). Industry 4.0 relies on adopting digital technologies to gather and analyze data in real time, providing useful information to the manufacturing system. The advent of the Internet of Things (IoT), cloud services, big data, and analytics made this possible, creating the cyber-physical system concept of Industry 4.0 (Munirathinam 2020). Research has been conducted on various Industry 4.0 technologies, such as artificial intelligence (AI), the Internet of Things (IoT), big data, and blockchain, and how they could create significant interruptions in recent years. These technologies provide various possibilities in the manufacturing and supply chain (Javaid et al. 2021b). Industry 4.0 technologies play an essential role in developing the smart supply chain. With predictive analytics methods, organizations could condense forecasting errors to nearly half. Industry 4.0 technologies can fulfill the demand of the medical supply chain during the COVID-19 pandemic. These are helpful for smartly tracking products during the COVID-19 pandemic. Industry 4.0 technologies are helpful for the effective management of the smart supply chain. These track inventories for smooth manufacturing processes (Verma et al. 2022).

The Fourth Industrial Revolution, called Industry 4.0, has the potential to influence the product design and development process (Bai et al. 2020). Integrating production systems and advanced design software in smart factories can enhance product design, development, and production processes through visualization and operations simulations (Lampropoulos and Siakas 2023). From the standpoint of Industry 4.0, promoting transparency and integration can facilitate heightened flexibility and prompt execution of personalized customer demands. Technological advancements, such as cyber-physical systems (CPSs) and the Internet of Things (IoT), have the potential to enhance visibility and connectivity within manufacturing and supply chain operations (Fatorachian and Kazemi 2021). Facilitating the instantaneous exchange of information and collaboration can facilitate rapid adaptability and customization of products to suit specific or regional requirements. Sharing information in real time can facilitate the creation of a collaborative production planning and control design, ultimately leading to product quality assurance (Fatorachian and Kazemi 2021).

Sophisticated machine learning enhances demand forecasting by analyzing diverse data sources. Probability distributions replace single projections, aiding supply chain decision-making. Predictive analytics reduces forecasting inaccuracies by 50%, optimizing inventory management for cost savings (Kulkarni et al. 2022).

The convergence of the Internet of Things and cyber-physical systems is observed. In the manufacturing sector, it is feasible for intelligent machines to collaborate with human workers to facilitate uncomplicated assembly operations (Suvarna et al. 2021). In addition, they can function independently and establish direct communication with production systems, thereby harnessing the potential of the Internet of Things. This enables them to resolve intricate issues and exercise autonomous decision-making (Moufaddal et al. 2020). These implications will significantly impact manufacturing. Advanced techniques enable cost-efficient custom orders. Implementation leads to reduced expenses and time in product design and engineering. Additionally, it will encourage the utilization of simulation and virtual testing at various stages of the product lifecycle (Okwu 2020).

Industry 5.0 is commonly recognized as the forthcoming industrial revolution. This approach aims to optimize manufacturing solutions by combining the ingenuity of human experts with advanced, intelligent, and precise machines, resulting in resource-efficient and user-centric outcomes that surpass those of Industry 4.0 (Aslam et al. 2020). Industry 5.0 is anticipated to benefit from various innovative technologies and applications that will facilitate enhanced production and the prompt delivery of personalized products (Maddikunta et al. 2022). Most companies face challenges in digitalizing their operations by incorporating advanced technologies such as artificial intelligence (AI), the Internet of Things (IoT), and cloud computing. However, it is anticipated that the next phase of the Industrial Revolution will soon emerge (Xu et al. 2021). Integrating evolving technology and changing human processes is expected to propel Industry 5.0 into the forefront of daily business operations.

The fifth Industrial Revolution holds significant potential for the strategic optimization of supply chains. The concept of Industry 5.0 aims to integrate the advanced features of highly automated, interconnected, and intelligent digital ecosystems with the human element (Huang et al. 2022). This entails utilizing the human factor to tailor end-user experiences, streamline workflows, and achieve optimal outcomes by effectively harnessing the capabilities of both personnel and technology involved in the process. Integrating cognitive computing and intelligent automation capabilities will facilitate hyperpersonalization, as human intelligence is empowered to work with these technologies (Leng et al. 2022). Utilizing computing techniques such as machine learning and robotic process automation can enhance the proficiency of front-line employees, resulting in the delivery of greater customer value in a more efficient manner. An experienced staff member can set up a complicated system for managing orders, enabling a less experienced employee to independently manage routine cases and prompt a request for assistance or instruction when faced with infrequent yet more demanding scenarios (Akundi et al. 2022). The integration of various technologies, such as augmented reality, 5G mobile, "cobots," automated drones, next-generation AI, and working exoskeletons, among others, has the potential to combine human vision, insights, and creativity with the increased efficiency of technology (Neights 2020).

A digital twin refers to a virtual representation of a tangible supply chain. Utilizing a virtual environment enables organizations to simulate their complete supply chain, test various scenarios, model different nodes, modes, flows, and policies, and assess the impact of decisions and disruptions on network operations (Empl and Pernul 2023). Utilizing a digital twin facilitates the ability of entities to conduct rigorous evaluations of their supply chain architecture. A crucial aspect of supply chain management is to develop a resilient supply chain capable of promptly recovering from disruptive events such as obstructed transportation routes and distribution center closures. Organizations utilize digital twins to enhance visibility throughout their supply chain, detect patterns, identify opportunities for improvement, eliminate inefficiencies, and optimize existing processes (Kamble et al. 2022). Digital twins play a crucial role in enhancing the resilience of supply chains by facilitating informed and prompt data-driven decision-making, enabling organizations to respond proactively rather than reactively in the event of any disruptions. Implementing a digital twin facilitates developing contingency strategies, determining optimal safety stock levels, identifying alternative transportation and logistics routes, and other related measures for organizational planning (Bhandal et al. 2022). The tool enables the execution of multiple scenarios concurrently, facilitating the resolution of distinct inquiries in various strategic domains. This approach ultimately leads to a conclusion that optimizes the business's overall performance (Buck 2022).

The emergence of Industry 5.0 is expected to necessitate the development of novel technologies, thereby mandating the need for commensurate progress in cyber security to counteract potential risks effectively. The advent of artificial intelligence is expected to usher in the next phase of progress in cyber security (Xian et al. 2023). Artificial intelligence has the potential to identify potential security risks, acquire knowledge of their patterns, and preemptively contain them before their infiltration. This will also result in novel and inventive methods for luring and neutralizing potential dangers. According to Ash (2019) findings, the system does not simply deny access to detecting a potential threat. Rather, it directs the threat toward a complex and intricate pathway that ultimately leads to no resolution.

The integration of supply chains is crucial to commercial and industrial operations. Numerous corporations rely on supply chains to manufacture goods or services for their clientele. Utilizing digital tools by supply chain entities can aid in optimizing their operational approaches, leading to heightened efficiency and a competitive edge. Moreover, implementing forthcoming intelligent supply chain technology can confer advantages to enterprises in this sector, provided that supply chain managers acquire a comprehensive outlook on these instruments (Adia 2022).

On average, leading organizations allocate a budget of $17 million toward integrating artificial intelligence in their supply chain management operations. The quantity above is expected to escalate, and more enterprises will adopt digital technology as business executives develop confidence in it and recognize its worth. The implementation of end-to-end automation and optimization of the value stream supply chain is imperative for the survival of businesses. According to Ballapragada (2019), there is a growing trend toward digitizing complex manufacturing operations management, leading to a greater reliance on robotics-driven functional aspects within organizations.

Predictive analytics tools enable intelligent supply chains to acquire crucial data and insights. This can additionally assist individuals in overseeing their equipment and the current state of crucial tasks. Moreover, predictive analytics to monitor equipment and tasks can aid supply chain enterprises in various processes, such as guaranteeing punctual delivery and formulating strategic goals (Adia 2022).

Organizations will learn that efficient supply chain management entails possessing pertinent data rather than abundant data. Hence, entities will initiate investments in technologies that facilitate the consolidation and optimization of data enhancement endeavors (Richter et al. 2022). The users will learn to depend on filtering mechanisms and integrate current, conventional, and upcoming data sources to construct a unified and comprehensive representation of their intelligent supply chain management. According to (Ballapragada 2019), the cloud computing industry is projected to reach $410 billion by 2020. This growth is expected to lead to increased consolidation of cloud-based data lakes in the manufacturing and supply chain management sectors, facilitating the integration of data-at-rest and data-in-motion concepts. This, in turn, is expected to drive innovation and improve infrastructure parameters.

It is projected that within the upcoming five-year period, the number of physical devices connected to the Internet will increase from 4 to 30 billion. These factors will substantially affect the functioning of supply chains (Suntsova 2022). As previously mentioned, the vast amount of data obtained from various internal and external sources, including the industrial Internet of Things, can enhance and expedite decision-making processes across numerous industries and applications (Li et al. 2022). It can be inferred that future advanced supply chain technologies will automate decision-making processes to guarantee the overall health and performance of the end-to-end supply chain. The utilization of big data is expected to bring about a significant transformation in the decision-making process of businesses throughout the procurement-to-pay cycle as they move upstream. Organizations that rely on data will frequently utilize real-time analytics to obtain significant metrics that were previously imperceptible. The outcome of this study will provide novel avenues for enhancing procedures, monitoring key performance indicators, and establishing benchmarks in an automated manner (Cheng et al. 2022).

It is advantageous to contemplate the present patterns within this sector to understand the future trajectory of intelligent supply chains. The employment of automation and artificial intelligence is gaining momentum in intelligent supply chain management and operations. Furthermore, the emergence of COVID-19 in recent years has led to increased human safety concerns and revised protocols promoting adequate sanitation. The impact of such trends on the operations of intelligent supply chains has resulted in significant changes, thereby influencing the outlook of supply chain management and future expectations (Adia 2022).

Real-time transportation visibility platforms employ software to enable supply chains to monitor the status of shipments during transit after they depart from the warehouse. Consequently, there is anticipation of an increase in demand for these commodities in the coming years. An increasing number of supply chains will probably adopt such tools to oversee their shipments and enhance the efficiency of their delivery procedures. Nonetheless, such an outcome is contingent upon the concurrence of platform vendors with intelligent chain carrier networks (Adia 2022).

According to Gartner's forecast, by 2024, approximately 50% of supply chain entities will have adopted software applications that facilitate the utilization of artificial intelligence and sophisticated analytical techniques. Supply chain leaders could adopt a comprehensive viewpoint on AI and advanced analytics to enhance efficiency across various supply chain management and operations domains. Adia posits that the features above in supply chain technology and tools are poised to experience a surge in popularity, potentially surpassing the growth rate projected by Gartner. The statement above pertains to the proliferation of proficient devices that endorse artificial intelligence and analytical capabilities in the current market. It is posited that this trend will gain greater acceptance in the coming years, as per Adia (2022) assertion.

Edges refer to the tangible sites within the supply chain network where various entities, including individuals, data connections, machines, sensors, and devices, are situated. Edge ecosystems facilitate proximity between the decision-making process and the information source. Edge ecosystems provide the necessary infrastructure for the seamless integration of automated network tools, devices, and applications, including but not limited to drones, robots, and connected vehicles. The progression of data communication services, such as Wi-Fi, Bluetooth, and 5G, will provide additional backing to edge ecosystems and supplement conventional centralized supply chain solutions. Edge computing decision-making is presently taking place across numerous supply chains. The forthcoming three years will be dedicated to identifying diverse use cases to facilitate connected, automated, and autonomous networks of edge decisions (Hippold 2022).

As the complexity and volatility of supply chains continue to rise, supply chain organizations must enhance their agility. This implies that conventional applications developed on ancient architectures are no longer suitable. To ensure the longevity of the technological foundation of the supply chain, a viable approach is to transition toward application architectures that are based on microservices and are compostable. Nonetheless, for the majority of organizations, the process of upgrading or implementing a new SCE application is both expensive and time-consuming. Supply chain executives frequently weigh the potential drawbacks of refraining from upgrades against the objective of minimizing expenses. Adopting modular and composable services in the supply chain is expected to occur early, particularly for enterprise-centric applications that can be internally managed. At a multi-enterprise level that involves diverse business partners and systems, challenges persist in achieving widespread adoption, as noted by (Hippold 2022).

8.4 Discussion and Implications

Data privacy and security are among the primary ethical concerns associated with the technological revolution in supply chain management. As organizations increasingly rely on data to optimize their supply chain operations, they must ensure that they are safeguarding sensitive information about their customers, employees, and partners. Failure to protect these data can result in significant reputational damage, legal liabilities, and financial losses. Another ethical concern is the potential for technology to exacerbate existing inequalities in the supply chain, particularly in terms of labor rights and environmental sustainability.

Adopting automation and artificial intelligence can potentially displace workers in the supply chain, particularly those in low-skilled and manual labor roles. Additionally, the increased reliance on technology can lead to overconsumption and waste, which can have negative environmental impacts. Moreover, there are concerns around the use of technologies such as blockchain, which can potentially create more transparency in the supply chain but can also be used to obscure unethical or illegal practices. There is also a risk that technologies such as predictive analytics can be used to discriminate against certain groups, particularly in areas such as hiring and promotion. First, business managers and entrepreneurs should prioritize data privacy and security to safeguard sensitive information about their customers, employees, and partners. This can be achieved by adopting robust data security protocols, employee training, and ongoing monitoring to detect potential breaches or vulnerabilities. Second, business managers and entrepreneurs should be mindful of technology's potential to exacerbate supply chain inequalities. This includes taking steps to mitigate the impact of automation on workers, such as reskilling and retraining programs, and ensuring that environmental sustainability is prioritized in supply chain operations. Third, business managers and entrepreneurs should consider the ethical implications of the technologies they adopt in the supply chain, such as blockchain and predictive analytics. This includes engaging in transparent and responsible practices that promote accountability and fairness. Business managers and entrepreneurs should also prioritize collaboration and partnership with their supply chain stakeholders, including suppliers, distributors, and logistics providers. This can help to ensure that everyone in the supply chain is working toward common goals and that ethical considerations are at the forefront of decision-making. Finally, business managers and entrepreneurs should maintain a commitment to ongoing learning and innovation, remaining up-to-date with the latest technological advancements and trends in supply chain management (Zaman et al. 2024a, b). This can help them identify improvement opportunities and stay ahead of the competition in an ever-evolving landscape.

First, academia and education institutions should prioritize research and development in supply chain management to keep up with the latest technological advancements and trends. This includes investing in interdisciplinary research and collaboration with industry partners to stay ahead of the curve. Second, academia and educational institutions should incorporate training and education on ethical considerations and social responsibility into their curricula. This can include courses on data privacy and security, environmental sustainability, and labor practices in the supply chain. Third, academia and educational institutions should prioritize developing and promoting diversity and inclusion in supply chain education and research. This includes promoting the participation of women, people of color, and other underrepresented groups in supply chain education and research and incorporating diverse perspectives into supply chain decision-making. Additionally, academia and education institutions should prioritize industry partnerships and collaboration to ensure that their research and education are relevant to the needs of the supply chain industry. This can involve establishing partnerships with industry leaders to provide students with practical experience and exposure to real-world supply chain challenges. Finally, academia and education institutions should maintain a commitment to ongoing learning and innovation, remaining up-to-date with the latest technological advancements and trends in supply chain management. This can help to ensure that their education and research remain relevant and impactful in an ever-evolving landscape.

Policymakers should promote and incentivize the adoption of sustainable and environmentally friendly practices in supply chain management. This can include implementing policies and regulations that require companies to disclose their environmental impact and reduce their carbon footprint. Second, policymakers should prioritize data privacy and security in supply chain management. This can involve establishing regulations that protect consumer data and require companies to be transparent about their data collection and sharing practices. Third, policymakers should support the development and adoption of emerging technologies such as blockchain, artificial intelligence, and the Internet of Things to improve the efficiency and transparency of the supply chain. This can include funding and resources to support research and development in these areas and creating policies encouraging companies to invest in these technologies. Additionally, policymakers should prioritize the development of a skilled workforce that is equipped to manage and operate the latest technologies in supply chain management. This can include funding education and training programs that teach the necessary skills and knowledge to work in the field. Finally, policymakers should encourage international collaboration and cooperation to address global supply chain challenges. This can involve working with other countries to establish common standards and regulations that promote ethical and sustainable supply chain practices.

8.5 Conclusion

As the technological revolution continues to unfold, the future of supply chain management is becoming increasingly dependent on digitization, artificial intelligence, and automation. These key technologies rapidly transform the industry, rewriting how companies source, purchase, produce, and deliver goods to customers. With the introduction of digitization, for example, companies can now gather and analyze data more meaningfully, making more informed decisions and responding more quickly to changes in demand and market conditions. Artificial intelligence is also transforming how companies operate, enabling them to automate workflows and optimize operations in previously impossible ways. With the rise of automation, companies can now streamline their supply chain processes, reducing costs and increasing efficiency levels. These technological advancements and increased customer expectations drive companies to find innovative solutions to optimize their supply chains and improve their overall performance. To remain competitive in this rapidly evolving industry, companies must embrace these changes and adapt their operations to take advantage of new technologies and opportunities (Zaman et al. 2024a, b). Of course, there may be challenges and potential disruptions as the industry adapts to these changes. Some jobs may be lost as companies automate certain tasks, and there could be short-term disruptions to supply chains.

The supply chain industry is undergoing a major transformation due to the rapid advancement of technology. With the help of cutting-edge technologies such as IoT sensors, blockchain, digital twins, edge computing, and AI, companies can improve their supply chain operations, reduce costs, and enhance customer satisfaction. However, this digital revolution comes with challenges, such as cybersecurity threats, data integration issues, and the need for skilled talent to manage new technologies. Companies must embrace digital transformation and invest in the latest technologies to keep pace with the ever-changing technological landscape. This means building a tech-savvy workforce, implementing agile supply chain systems, and partnering with suppliers, customers, and other stakeholders to leverage data and insights. In addition to investing in technology, collaboration among industry stakeholders and implementing global standards can help enhance supply chain visibility, sustainability, and resilience. Companies need to take a holistic approach to their supply chain operations, considering the entire supply chain ecosystem to identify opportunities for optimization and efficiency.

Furthermore, sustainability is becoming increasingly important to customers, and companies need to integrate sustainable practices into their supply chain operations to remain competitive in the market. Sustainable practices can reduce costs, enhance reputation, and attract environmentally conscious customers. In conclusion, the future of supply chain management is inextricably linked to technological innovation. Companies that can successfully leverage technology to optimize their supply chain operations while meeting changing customer demands and implementing sustainable practices will be well-positioned to succeed. Companies that fail to adapt will likely fall behind their competitors and struggle to survive in the ever-changing supply chain.

Acknowledgements In the preparation of this manuscript, we utilized AI tools to assist with various tasks, including language editing and content organization. The use of AI was conducted within ethical boundaries, ensuring that it complemented our research and writing efforts without compromising the originality or integrity of our work. We acknowledge and appreciate the support provided by these technologies, which have contributed to the clarity and readability of the chapter.

References

Aamer, A., Sahara, C. R., & Al-Awlaqi, M. A. (2023). Digitalization of the supply chain: transformation factors. *Journal of Science and Technology Policy Management, 14*(4), 713–733.

Adia. (2022). The Future of Smart Supply Chains. Retrieved from https://adia.works/blog/smart-supply-chain/.

Aeologic. (2022). The Impact of Technology on Supply Chain Management. Retrieved from https://www.aeologic.com/blog/the-impact-of-technology-on-supply-chain-management/.

Agarwal, U., Rishiwal, V., Tanwar, S., Chaudhary, R., Sharma, G., Bokoro, P. N., & Sharma, R. (2022). Blockchain technology for secure supply chain management: A comprehensive review. *IEEE Access*.

Ajayi, M. O., & Laseinde, O. T. (2023). A review of supply chain 4IR management strategy for appraising the manufacturing industry's potentials and shortfalls in the 21st century. *Procedia computer science, 217*, 513–525.

Akundi, A., Euresti, D., Luna, S., Ankobiah, W., Lopes, A., & Edinbarough, I. (2022). State of Industry 5.0—Analysis and identification of current research trends. *Applied System Innovation, 5*(1), 27.

Amesite. (2021). The Evolution of Industry from 1.0 to 4.0. Retrieved from https://amesite.com/blogs/the-evolution-of-industry-from-1-0-to-4-0/.

Ammar, M., Haleem, A., Javaid, M., Bahl, S., & Nandan, D. (2022). Improving the performance of supply chain through industry 4.0 technologies. In *Advancement in Materials, Manufacturing and Energy Engineering, Vol. II: Select Proceedings of ICAMME 2021* (pp. 197–209): Springer.

Ancillai, C., Sabatini, A., Gatti, M., & Perna, A. (2023). Digital technology and business model innovation: A systematic literature review and future research agenda. *Technological Forecasting and Social Change, 188*, 122307.

Andronie, M., Lăzăroiu, G., Ştefănescu, R., Uţă, C., & Dijmărescu, I. (2021). Sustainable, smart, and sensing technologies for cyber-physical manufacturing systems: A systematic literature review. *Sustainability, 13*(10), 5495.

Ash, B. (2019). Introducing Industry 5.0 the new-new and how it will reinvent manufacturing. Retrieved from https://www.linkedin.com/pulse/introducing-industry-50-new-new-how-reinvent-bernard-ash/.

Aslam, F., Aimin, W., Li, M., & Ur Rehman, K. (2020). Innovation in the era of IoT and industry 5.0: Absolute innovation management (AIM) framework. *Information, 11*(2), 124.

Bai, C., Dallasega, P., Orzes, G., & Sarkis, J. (2020). Industry 4.0 technologies assessment: A sustainability perspective. *International journal of production economics, 229*, 107776.

Ballapragada, B. (2019). Smart Supply Chain Management: Why it is Important to Implement? Retrieved from https://throughput.world/blog/smart-supply-chain-management/.

Bhandal, R., Meriton, R., Kavanagh, R. E., & Brown, A. (2022). The application of digital twin technology in operations and supply chain management: a bibliometric review. *Supply Chain Management: An International Journal, 27*(2), 182–206.

Bhattacharyya, S. S., Maitra, D., & Deb, S. (2021). Study of adoption and absorption of emerging technologies for smart supply chain management: a dynamic capabilities perspective. *International Journal of Applied Logistics (IJAL), 11*(2), 14–54.

Buck, S. (2022). The Complete Guide To Supply Chain Digital Twins. Retrieved from https://www.coupa.com/blog/supply-chain/complete-guide-supply-chain-digital-twins.

Bühler, M. M., Jelinek, T., & Nübel, K. (2022). Training and preparing tomorrow's workforce for the fourth industrial revolution. *Education Sciences, 12*(11), 782.

Burrus, D. (2022). Industry 4.0 technologies. Retrieved from https://www.sap.com/mena/insights/what-is-industry-4-0.html#:~:text=Industry%204.0%20is%20powered%20by,machinery%2C%20robots%2C%20and%20vehicles.

Butt, R., Siddiqui, H., Soomro, R. A., & Asad, M. M. (2020). Integration of Industrial Revolution 4.0 and IOTs in academia: a state-of-the-art review on the concept of Education 4.0 in Pakistan. *Interactive Technology and Smart Education, 17*(4), 337–354.

Centobelli, P., Cerchione, R., & Ertz, M. (2020). Agile supply chain management: where did it come from and where will it go in the era of digital transformation? *Industrial Marketing Management, 90*, 324–345.

Davydov, R. (2022). A guide to making your supply chain smart. Retrieved from https://www.itransition.com/blog/smart-supply-chain.

Dekavadiya, A. (2021). How The Manufacturing Industry Evolved: Industrial Revolution 1.0 to 4.0. Retrieved from https://www.aimdek.com/how-the-manufacturing-industry-evolved-industrial-revolution-1-0-to-4-0/.

Desoutter. (2022). Industrial Revolution—From Industry 1.0 to Industry 4.0. Retrieved from https://www.desouttertools.com/your-industry/news/503/industrial-revolution-from-industry-1-0-to-industry-4-0.

DiConneX. (2020). The Industrial Revolution—from Industry 1.0 to Industry 4.0. Retrieved from https://diconnex.com/en/blog/2020/06/23/the-industrial-revolution/.

Edwin Cheng, T., Kamble, S. S., Belhadi, A., Ndubisi, N. O., Lai, K.-h., & Kharat, M. G. (2022). Linkages between big data analytics, circular economy, sustainable supply chain flexibility, and sustainable performance in manufacturing firms. *International Journal of Production Research, 60*(22), 6908–6922.

Ellitan, L. (2020). Competing in the era of industrial revolution 4.0 and society 5.0. *Jurnal Maksipreneur: Manajemen, Koperasi, dan Entrepreneurship, 10*(1), 1–12.

Empl, P., & Pernul, G. (2023). Digital-Twin-Based Security Analytics for the Internet of Things. *Information, 14*(2), 95.

Fatorachian, H., & Kazemi, H. (2021). Impact of Industry 4.0 on supply chain performance. *Production Planning & Control, 32*(1), 63–81.

Fekri, M., Jafari, M., & Mahdavi Mazdeh, M. (2022). The Impact of Industry 4.0 on the Automotive Business Models: A Successful Transition Through an Integrated System-Engineering and Strategic Perspective. *Available at SSRN 4181782.*

Frauendorf, J. L., & de Souza, É. A. (2023). *The Architectural and Technological Revolution of 5G*: Springer.

Geographic, N. (2022). Industrial Revolution and Technology. Retrieved from https://education.nationalgeographic.org/resource/industrial-revolution-and-technology/.

Har, L. L., Rashid, U. K., Te Chuan, L., Sen, S. C., & Xia, L. Y. (2022). Revolution of retail industry: from perspective of retail 1.0 to 4.0. *Procedia computer science, 200*, 1615–1625.

Hayes, A. (2022). The Supply Chain: From Raw Materials to Order Fulfillment. Retrieved from https://www.investopedia.com/terms/s/supplychain.asp.

Hernandez-de-Menendez, M., Morales-Menendez, R., Escobar, C. A., & McGovern, M. (2020). Competencies for industry 4.0. *International Journal on Interactive Design and Manufacturing (IJIDeM), 14*, 1511–1524.

Hippold, S. (2022). Gartner Predicts the Future of Supply Chain Technology. Retrieved from https://www.gartner.com/smarterwithgartner/gartner-predicts-the-future-of-supply-chain-technology.

Hu, H., Xu, J., Liu, M., & Lim, M. K. (2023). Vaccine supply chain management: An intelligent system utilizing blockchain, IoT and machine learning. *Journal of business research, 156*, 113480.

Huang, S., Wang, B., Li, X., Zheng, P., Mourtzis, D., & Wang, L. (2022). Industry 5.0 and Society 5.0—Comparison, complementation and co-evolution. *Journal of Manufacturing Systems, 64*, 424–428.

Javaid, M., Haleem, A., Singh, R. P., & Suman, R. (2021). Substantial capabilities of robotics in enhancing industry 4.0 implementation. *Cognitive Robotics, 1*, 58–75.

Javaid, M., Haleem, A., Singh, R. P., Khan, S., & Suman, R. (2021). Blockchain technology applications for Industry 4.0: A literature-based review. *Blockchain: Research and Applications, 2*(4), 100027.

Javaid, M., Haleem, A., Singh, R. P., & Suman, R. (2022). Artificial intelligence applications for industry 4.0: A literature-based study. *Journal of Industrial Integration and Management, 7*(01), 83–111.

Jiang, Z., Yuan, S., Ma, J., & Wang, Q. (2022). The evolution of production scheduling from Industry 3.0 through Industry 4.0. *International Journal of Production Research, 60*(11), 3534–3554.

Kamble, S. S., Gunasekaran, A., Parekh, H., Mani, V., Belhadi, A., & Sharma, R. (2022). Digital twin for sustainable manufacturing supply chains: Current trends, future perspectives, and an implementation framework. *Technological Forecasting and Social Change, 176*, 121448.

Kar, A. K., Choudhary, S. K., & Singh, V. K. (2022). How can artificial intelligence impact sustainability: A systematic literature review. *Journal of Cleaner Production*, 134120.

Kulkarni, V., Barat, S., Clark, T., & Barn, B. S. (2022). *Digital twin as an aid for decision-making in the face of uncertainty.* Paper presented at the 2022 Winter Simulation Conference (WSC).

Kumar, A., Agrawal, R., Wankhede, V. A., Sharma, M., & Mulat-Weldemeskel, E. (2022). A framework for assessing social acceptability of industry 4.0 technologies for the development of digital manufacturing. *Technological Forecasting and Social Change, 174*, 121217.

Lampropoulos, G., & Siakas, K. (2023). Enhancing and securing cyber-physical systems and Industry 4.0 through digital twins: A critical review. *Journal of Software: Evolution and Process, 35*(7), e2494.

Lazaroiu, G., Androniceanu, A., Grecu, I., Grecu, G., & Neguriță, O. (2022). Artificial intelligence-based decision-making algorithms, Internet of Things sensing networks, and sustainable cyber-physical management systems in big data-driven cognitive manufacturing. *Oeconomia Copernicana, 13*(4), 1047–1080.

Lee, K., Azmi, N., Hanaysha, J., Alzoubi, H., & Alshurideh, M. (2022). The effect of digital supply chain on organizational performance: An empirical study in Malaysia manufacturing industry. *Uncertain Supply Chain Management, 10*(2), 495–510.

Leng, J., Sha, W., Wang, B., Zheng, P., Zhuang, C., Liu, Q., ... Wang, L. (2022). Industry 5.0: Prospect and retrospect. *Journal of Manufacturing Systems, 65*, 279–295.

Li, C., Chen, Y., & Shang, Y. (2022). A review of industrial big data for decision making in intelligent manufacturing. *Engineering Science and Technology, an International Journal, 29*, 101021.

Luo, W., & Lin, J. (2021). *Research on improvement of computer software technology based on internet technology.* Paper presented at the Journal of Physics: Conference Series.

Maddikunta, P. K. R., Pham, Q.-V., Prabadevi, B., Deepa, N., Dev, K., Gadekallu, T. R., ... Liyanage, M. (2022). Industry 5.0: A survey on enabling technologies and potential applications. *Journal of Industrial Information Integration, 26*, 100257.

Melnyk, L. H., Matsenko, O. M., Kalinichenko, L. L., Holub, A. V., & Sotnyk, I. M. (2023). Instruments for ensuring the phase transition of economic systems to management based on Industries 3.0, 4.0, 5.0.

Microsoft. (2022). What is Industry 4.0? Retrieved from https://www.microsoft.com/en-us/industry/manufacturing/resources/industry-4-0-technologies.

Mohammed, A., Kannan, G., Zubairu, N., Pratabaraj, J., & Abideen, A. Z. (2023). Multi-tier supply chain network design: A key towards sustainability and resilience. *Computers & Industrial Engineering*, 109396.

Mosteiro-Sanchez, A., Barcelo, M., Astorga, J., & Urbieta, A. (2022). End to end secure data exchange in value chains with dynamic policy updates. arXiv preprint arXiv:2201.06335.

Moufaddal, M., Benghabrit, A., & Bouhaddou, I. (2020). *A cyber-physical warehouse management system architecture in an Industry 4.0 context.* Paper presented at the International Conference on Artificial Intelligence & Industrial Applications.

Muhammad, M. S., Kerbache, L., & Elomri, A. (2022). *Potential of additive manufacturing for upstream automotive supply chains.* Paper presented at the Supply Chain Forum: An International Journal.

Munirathinam, S. (2020). Industry 4.0: Industrial internet of things (IIOT). In *Advances in computers* (Vol. 117, pp. 129–164): Elsevier.

Neights, G. (2020). Industry 5.0 And The Supply Chain. Retrieved from https://talkinglogistics.com/2020/08/11/industry-5-0-supply-chain/.

nibusinessinfo. (2022). Industry 4.0: What it means for your business. Retrieved from https://www.nibusinessinfo.co.uk/content/industry-4.0-technologies.

Okwu, M. (2020). A review of the effect of Industry 4.0 on supply chain systems. Retrieved from https://www.researchgate.net/publication/354555200_A_review_of_the_effect_of_Industry_40_on_supply_chain_systems.

Özdağoğlu, A., & Bahar, S. (2022). Logistics 4.0 and smart supply chain management. In *Industry 4.0 and Global Businesses* (pp. 163–183): Emerald Publishing Limited.

Park, A., & Li, H. (2021). The effect of blockchain technology on supply chain sustainability performances. *Sustainability, 13*(4), 1726.

Rad, F. F., Oghazi, P., Palmié, M., Chirumalla, K., Pashkevich, N., Patel, P. C., & Sattari, S. (2022). Industry 4.0 and supply chain performance: A systematic literature review of the benefits, challenges, and critical success factors of 11 core technologies. *Industrial Marketing Management, 105*, 268–293.

Raja Santhi, A., & Muthuswamy, P. (2023). Industry 5.0 or industry 4.0 S? Introduction to industry 4.0 and a peek into the prospective industry 5.0 technologies. *International Journal on Interactive Design and Manufacturing (IJIDeM), 17*(2), 947–979.

Raji, I. O., Shevtshenko, E., Rossi, T., & Strozzi, F. (2021). Industry 4.0 technologies as enablers of lean and agile supply chain strategies: an exploratory investigation. *The International Journal of Logistics Management.*

Rawat, R. (2022). A Systematic Review of Blockchain Technology Use in E-Supply Chain in Internet of Medical Things (Iomt). *International Journal of Computations, Information and Manufacturing (IJCIM), 2*(2).

Richter, L., Lehna, M., Marchand, S., Scholz, C., Dreher, A., Klaiber, S., & Lenk, S. (2022). Artificial intelligence for electricity supply chain automation. *Renewable and Sustainable Energy Reviews, 163*, 112459.

Rodríguez-Espíndola, O., Chowdhury, S., Beltagui, A., & Albores, P. (2020). The potential of emergent disruptive technologies for humanitarian supply chains: The integration of blockchain, artificial intelligence and 3D printing. *International Journal of Production Research, 58*(15), 4610–4630.

Rosairo, P. (2021). THE SMART SUPPLY CHAIN STUFF IS REALLY TRICKY. Retrieved from https://ciltinternational.org/education-development/publications-articles/publication/the-smart-supply-chain-stuff-is-really-tricky/.

Rupp, M., Schneckenburger, M., Merkel, M., Börret, R., & Harrison, D. K. (2021). Industry 4.0: A technological-oriented definition based on bibliometric analysis and literature review. *Journal of Open Innovation: Technology, Market, and Complexity, 7*(1), 68.

Sahoo, S., & Lo, C.-Y. (2022). Smart manufacturing powered by recent technological advancements: A review. *Journal of Manufacturing Systems, 64*, 236–250.

Sharma, V., Raut, R. D., Hajiaghaei-Keshteli, M., Narkhede, B. E., Gokhale, R., & Priyadarshinee, P. (2022). Mediating effect of industry 4.0 technologies on the supply chain management practices and supply chain performance. *Journal of Environmental Management, 322*, 115945.

Sime. (2021). Industrial Revolution 1.0—Era of Mechanization. Retrieved from https://medium.com/spark-by-sime/industrial-revolution-1-0-9e6dc9c62c8c.

Simon. (2023). Industry 1.0 To 4.0—Brief History of The Industrial Revolution. Retrieved from https://sustainability-success.com/industry-1-0-to-4-0-2-3-revolution/.

Suntsova, O. (2022). The definition of smart economy and digital transformation of business in the concepts Industry 4.0 and 5.0. *Technology audit and production reserves, 4*(4/66), 18–23.

Suvarna, M., Yap, K. S., Yang, W., Li, J., Ng, Y. T., & Wang, X. (2021). Cyber–physical production systems for data-driven, decentralized, and secure manufacturing—A perspective. *Engineering, 7*(9), 1212–1223.

Tallera, S. (2022). Industrial Revolution 1.0 to 4.0—From Steam to Sensors. Retrieved from https://www.linkedin.com/pulse/industrial-revolution-10-40-from-steam-sensors-sandesh-tallera/?trk=articles_directory.

Verma, P., Samuel, O. D., Verma, T. N., & Dwivedi, G. (2022). *Advancement in Materials, Manufacturing and Energy Engineering, Vol. II*: Springer.

Vinitha, K., Prabhu, R. A., Bhaskar, R., & Hariharan, R. (2020). Review on industrial mathematics and materials at Industry 1.0 to Industry 4.0. *Materials Today: Proceedings, 33*, 3956–3960.

Wiechmann, D. M., Reichstein, C., Haerting, R.-C., Bueechl, J., & Pressl, M. (2022). Agile management to secure competitiveness in times of digital transformation in medium-sized businesses. *Procedia computer science, 207*, 2353–2363.

Wizata. (2022). TECHNOLOGIES BEHIND INDUSTRIAL REVOLUTION 4.0. Retrieved from https://www.wizata.com/knowledge-base/technologies-behind-industrial-revolution-4.0.

Xian, W., Yu, K., Han, F., Fang, L., He, D., & Han, Q.-L. (2023). Advanced Manufacturing in Industry 5.0: A Survey of Key Enabling Technologies and Future Trends. *IEEE Transactions on Industrial Informatics*.

Xu, X., Lu, Y., Vogel-Heuser, B., & Wang, L. (2021). Industry 4.0 and Industry 5.0—Inception, conception and perception. *Journal of Manufacturing Systems, 61*, 530–535.

Yang, M., Fu, M., & Zhang, Z. (2021). The adoption of digital technologies in supply chains: Drivers, process and impact. *Technological Forecasting and Social Change, 169*, 120795. https://doi.org/10.1016/j.techfore.2021.120795.

Zaman, S. A. A., Vilkas, M., & Zehra, F. T. (2023, 9–11 Oct. 2023). *Investment in digitalization, revenue enhancement through digitalization, and firm performance: Examining the moderating role of contingency factors1*. Paper presented at the 2023 IEEE International Conference on Technology and Entrepreneurship (ICTE).

Zaman, S. A. A., Yushi, J., Khan, S., Jamil, S., & Zaman, S. I. (2024). Knowledge Creation in SMEs in the Era of Industry 4.0: a Comparative Study of Pakistan and China. *Journal of the Knowledge Economy*. https://doi.org/10.1007/s13132-024-02248-6.

Zaman, S. A. A., Vilkas, M., Zaman, S. I., & Jamil, S. (2024). Digital technologies and digitalization performance: the mediating role of digitalization management. *JOURNAL OF MANUFACTURING TECHNOLOGY MANAGEMENT, ahead-of-print* (ahead-of-print). https://doi.org/10.1108/JMTM-04-2024-0176.

Zhang, G., Yang, Y., & Yang, G. (2023). Smart supply chain management in Industry 4.0: the review, research agenda and strategies in North America. *Annals of Operations Research, 322*(2), 1075–1117.

Zonar. (2023). INDUSTRIAL REVOLUTION 3.0. Retrieved from https://www.linkedin.com/pulse/industrial-revolution-30-zeekoi/.

Chapter 9
Managing Firms Supply Chain Performance: The Mediating Role of *Smart* Technologies

Sajid Nazir, Jyri Vilko, and Marko Torkkeli

9.1 Introduction

Supply chain management (SCM) comprises a series of activities and coordination among supply chain partners ensuring a smooth flow of product between upstream to downstream partners. This requires collaboration and interdependence among the different entities in the supply chain (Chopra 2020). Nevertheless, it also makes the supply chain vulnerable to risks arising in any part of the supply chain and impacting all its members (Ivanov 2021). The impact of risk and associated challenges can cost considerable losses in the form of time, effort, and finances (Xu et al. 2020; Ye et al. 2022). Thus, a failure at any stage can result in considerable losses throughout the supply chain (Ivanov et al. 2018). For example, a supplier's delay in Sony's PlayStation 3 shipments caused a decline in revenue and long-term market share (Mokhtar et al. 2017). In recent years, there have been several examples of negative impacts of supply chain disruptions which adversely affected the manufacturing industry in Pakistan as well as the global economy (Li and Zhang 2023). Such as, the emergence of COVID-19 negatively affected the global supply chain performance (Rapaccini et al. 2020; Shokouhifar and Ranjbarimesan 2022) affecting the product deliveries and shortages of raw material (Tietze et al. 2022; Nazir et al. 2024). As a result, a significant proportion (94%) of Fortune 1000 firms have encountered disruptions in their supply chain activities (Sherman 2020). Similarly, due to the

S. Nazir (✉) · J. Vilko · M. Torkkeli
Industrial Engineering and Management, Lappeenranta-Lahti University of Technology (LUT), Lappeenranta, Finland
e-mail: sajid.nazir@lut.fi

J. Vilko
e-mail: Jyri.Vilko@lut.fi

M. Torkkeli
e-mail: Marko.Torkkeli@lut.fi

M. S. Mubarik and S. A. Khan (eds.), *Smart Supply Chain Management*,
https://doi.org/10.1007/978-981-96-1333-5_9

Wuhan shutdown, certain Foxconn plants in China had to close, which caused Apple to postpone the release of new goods (Xu et al. 2020; Ye et al. 2022). These issues not only disrupt the company's ability to derive its own benefits, but they also have a direct impact on the overall supply chain performance (SCP) of an organization.

In order to address the damaging effects of the COVID-19 pandemic on business operations, it is essential for companies to strategically enhance their supply chain management practices (Fatorachian and Kazemi 2020; Vanteddu and Bandyopadhayay 2022). Such as the challenges faced by today's supply chain are becoming complex and diverse, demanding a better strategy to cope with them. One strategy that has found prominence in ensuring visibility across the supply chain process is the adoption of smart technologies (Fatorachian and Kazemi 2020; AlMulhim 2021; Vanteddu and Bandyopadhayay 2022). The utilization of smart technologies facilitates the advancement of supply chain management (Nasiri et al. 2020) and provides real-time information on the movement of goods and up-to-date information on the supply chain environment and expected challenges (AlMulhim 2021). The advent of smart technologies has the potential to develop a visible supply chain that is capable of sharing information and enabling proactive planning (Zhong et al. 2017; Modgil and Sakhuja 2022). Furthermore, the incorporation of smart technologies into the supply chain has the potential to integrate, communicate, and automate the processes in a supply chain with real-time information sharing and improved collaboration (Nazir et al. 2023).

Through the adoption of smart technologies, today's supply chain planners are in a better position to collect and analyze the data for improved forecast, planning, and coordination. In doing so, the digital transformation in an organization is considered as a significant asset that holds the potential to facilitate end-to-end transparency, forecast potential dangers, and mitigate demand uncertainty (Ivanov et al. 2018). Furthermore, it plays a crucial role in facilitating administrative processes by enabling the redesign or reconstruction of business models in response to improvements in the market system (Pramanik et al. 2019). According to Verhoef et al. (2021), digital transformation consists of three distinct phases namely (digitalization, digitization, and digital transformation). During the initial phase, the organization incorporates digital technologies in order to enhance and revolutionize its business processes. While in the second stage, distinct functions are employed, such as making decisions regarding the transition of retailers from a physical location to an online storefront. Whereas in the third phase, companies modify the value of their products through the implementation of digital technologies. The process of digital transformation not only enhances the core production process of firms, but also amplifies their capacity to gather data for various purposes, including supplementing sales, identifying emerging marketplaces, streamlining client engagement, etc. (Frank et al. 2019).

Therefore, in order to facilitate efficient sharing of information, organizations are allocating resources toward technology advancements aimed at establishing robust communication platforms and collaborative processes. This strategic investment is intended to increase supply chain performance by fostering improved information-sharing capabilities (Abbasnejad et al. 2020; Ali et al. 2023). In doing so, it helps in lowering the product processing time, improves customer value, and increases the firm's profitability by lowering transportation and labor costs. While in the absence of it, organizations may be susceptible to numerous risks such as delays in supplies (Elsersy et al. 2021), inconsistencies in business operations (Omar et al. 2022), reduced competitiveness (Ji et al. 2022), customer dissatisfaction (Gupta et al. 2022) and long-term supplier relationship failures (Raj et al. 2022). Thus, the adoption of technology-driven strategies enables business managers to stay competitive in a fast-changing industry. It enables supply chain visibility and transparency and allows firm managers to have greater control over business operations including supply chain performance (Govindan et al. 2017; Junaid et al. 2023).

Nowadays, global technologies, particularly technologically advanced products, have become an essential tool for businesses to maintain viable relationships and develop substantial relationships with their stakeholders (Saberi et al. 2018). Furthermore, these technologies incorporate distinctive embedded systems and techniques that enable the real-time monitoring of the supply chain operations and facilitate better customer satisfaction and the firm's performance (Zahid et al. 2023). According to Deloitte AG's report in 2015, the incorporation of cutting-edge technology into businesses is crucial for enhancing productivity, reducing operational costs, and achieving high levels of customer satisfaction (Schroeck et al. 2019). The positive impact of the application of these new smart technologies has started to show results across the globe in diverse fields (Fatorachian and Kazemi 2020; Nasiri et al. 2020). However, there is an inadequate level of studies examining the implications of smart technologies in improving supply chain performance, particularly in manufacturing firms. Considering the potential of these smart technologies in improving supply chain processes, this study aims to examine the influence of digital transformation on supply chain performance of manufacturing industry. Furthermore, this study also evaluates the mediating effect of smart technologies in the association between digital transformation and supply chain performance in the manufacturing industry.

Figure 9.1 represents the evolution of supply chain. This evolution period dates were derived from (Ivanov and Sokolov 2010) in the creation of this figure. This represents that how the innovations evolved overtime, highlighting the significant moments that have shaped the current technological landscape with respect to supply chain. In doing so, the remaining sections of this study are organized in the following manner. Section one explores the relevant literature and gap for each of the three main research areas: digital transformation, supply chain performance, and smart technologies. Followed by section one, the hypothesis development and conceptual framework has been discussed in section two, while the research methodology has been presented in section three. Moreover, the fourth section of the paper outlines the findings of the study, whereas the fifth section is dedicated to the discussions and interpretation of these results. Finally, the concluding section, section six, provides an overview of the limitations of the study and identifies areas for future research in this field.

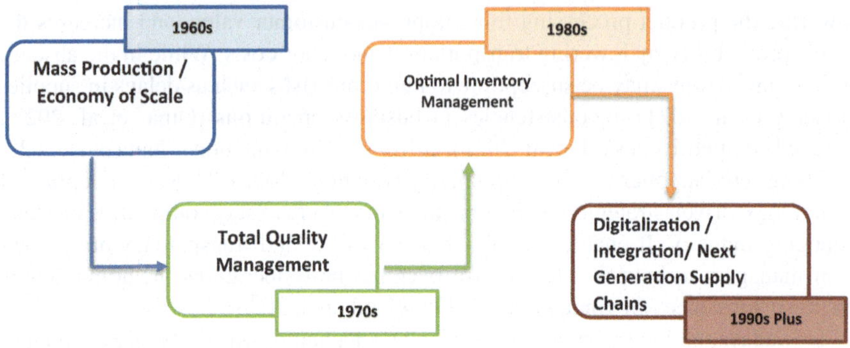

Fig. 9.1 Evolution of SCM (authors creation)

9.2 Literature Review

9.2.1 Definitions and Dimensions

This chapter focuses on three major facets that are critical to modern supply chains: digital transformation, smart technologies, and supply chain performance. Each of these elements plays a significant role in shaping the way supply chains operate in today's dynamic business environment. By exploring these concepts in depth, this chapter aims to provide a comprehensive understanding of how digital advancements and smart technologies contribute to enhancing overall supply chain performance.

(a) ***Smart technologies***: These technologies are particularly important in the context of smart supply chains, where the goal is to create an interconnected, real-time system that leverages advanced tools to optimize operations. In a smart supply chain, technologies such as IoT enable real-time tracking and monitoring, AI enhances predictive capabilities and planning, blockchain ensures transparency and security across the network, and digital twins simulate different supply chain scenarios for proactive problem-solving. These technologies make sense in the smart supply chain context because they provide the infrastructure for greater visibility, automation, and adaptability. By integrating these tools, companies can create a supply chain that is not only more efficient but also capable of self-optimizing and quickly responding to disruptions or changes in demand.

(b) ***Digital transformation***: It refers to the integration of digital technologies across all areas of a business, revolutionizing processes, operations, and customer interactions. In the context of supply chains, digital transformation involves transitioning from traditional, manual workflows to automated, data-driven processes that leverage tools such as IoT, AI, blockchain, and cloud computing. This transformation enables supply chains to become more agile, responsive, and data-informed, ultimately leading to greater efficiency, improved decision-making, and enhanced customer satisfaction.

(c) ***Supply chain performance***: SC performance is the ultimate measure of success, encompassing how effectively and efficiently a supply chain delivers products and services while minimizing costs and maximizing customer satisfaction. Supply chain performance is measured through metrics such as delivery times, cost reduction, order accuracy, and the ability to adapt to market shifts. With the integration of digital transformation and smart technologies, supply chain performance can be significantly enhanced, allowing for real-time visibility, better forecasting, and more streamlined operations. In sum, the convergence of these three facets—digital transformation, smart technologies, and supply chain performance—creates a robust framework for optimizing supply chain processes and meeting the evolving demands of today's global market.

In order to avoid any confusion, we have taken the below definitions of these constructs for the sake of this chapter:

Constructs definitions	
Digital transformation	"*Digital transformation refers to the process of introducing digital technology within an organization to foster innovation in business operations and production, hence enabling the organization to successfully meet demands* (Warner and Wäger 2019; Li 2020)"
Smart technologies	"*Smart technologies, comprising various devices and methods, serve as the fundamental building blocks for achieving effective digital transformation* (Yoo 2010; Nasiri et al. 2020)"
Supply chain performance	"*The measurement of supply chain performance is a complex aspect within the field of supply chain management. It encompasses various dimensions such as diversity, innovation, period of time, price, and availability* (Dreyer et al. 2016). *Additionally, the analysis of flexibility, reliability, responsiveness, quality, and asset management further contributes to the evaluation of supply chain performance* (Najmi and Makui 2011)"

9.2.2 Hypothesis Development

9.2.2.1 Digital Transformation and Smart Technologies

The concept of digitalization has changed how individuals deal with each other and their surroundings (Fabbe-Costes and Lechaptois 2022). The advent of novel technologies and devices, such as mobile phones, personal computers, driverless vehicles, and smart gadgets, has transformed how individuals' access and share information (Nasiri et al. 2020). As a result of technological advancements, manufacturing organizations are under more competitive pressure to improve quality, reduce costs, and speed up production schedules (Chen et al. 2020; Sucupira Furtado et al. 2023). Therefore, the widespread impact of novel technological advancements and digital transformations has been experienced across various sectors including supply chains

(Nasiri et al. 2020; Zhu 2022). The process of digitalization is expected to bring about significant transformations in supply chain management. Consequently, the integration of digital technology in supply chain management has broadened to incorporate both digital goods and services, as well as the management of all processes associated with supply chains (Attaran 2020). In doing so the primary aim of the digital supply chain is to enhance the awareness of systems regarding emerging developments and enable them to smartly modify the physical operations throughout the chain for maximum efficiency (Michel 2017). Nevertheless, the digital transformations of companies enable them to derive advantages from various advancements such as the use of barcodes, location-based services, GPS tracking, and wireless communications (Kearney 2015; AlMulhim 2021). However, all such activities are facilitated by smart technologies. Smart technologies can facilitate various activities by incorporating a set of features into technological devices that promote technical capabilities (Nasiri et al. 2020; Zhu 2022). Likewise, the use of smart technologies to improve coordination between various corporate operations is a primary objective of digital transformation (Berman 2012). Considering the impact of digital transformation on an organization's SC performance due to novel smart technologies, the following hypothesis has been proposed.

H1: *Digital transformation positively influenced by the adoption of smart technologies.*

9.2.2.2 Digital Transformation and Supply Chain Performance

The emergence of digital technologies has brought about a significant transformation in various conventional business processes (Nasiri et al. 2020; Zhu 2022; Vanteddu and Bandyopadhayay 2022) and has witnessed a steady rise over the past two decades (AlMulhim 2021). The implementation of digital transformation has resulted in the transition of companies' supply chains toward digital supply chain management, with the aim of enhancing customer service quality (Agrawal and Narain 2018; Verhoef et al. 2021). Therefore, companies are forced to adapt digital technologies in order to compete with the global market or be left behind (Chen et al. 2016; Büyüközkan and Göçer 2018; Chege et al. 2020). A considerable number of businesses are making significant investments in the digitalization of their business models, with a specific focus on enhancing their supply chain performance (Agrawal and Narain 2018; Fatorachian and Kazemi 2020). The management of supply chain functions is a crucial component of any business organization (Dreyer et al. 2016). As the never-ending escalation of supply chain expenses has a significant impact on the financial performance of all the parties involved; therefore, cutting costs is essential to staying competitive in today's global supply chain (Attaran 2020). In doing so, the utilization of advanced technologies to improve supply chain efficiencies can result in substantial cost savings for companies, particularly in transportation, holding inventory costs, and the overall supply chain performance of an organization (Fatorachian and Kazemi 2020; AlMulhim 2021). In addition to this, exceptional performance throughout the supply chain carries a strategic value that can result in a speedy return on investment (financial) and increases in both profitability and efficiency (Attaran 2012). Considering the impact of digital transformation on firms SC performance, the following hypothesis has been proposed.

H2: *Digital transformation positively influences the supply chain performance of any organization.*

9.2.2.3 Smart Technologies and Supply Chain Performance

The rapid development in technology adoption by a firm plays a significant role in improving the profitability of its business operations and enhance its capacity to mitigate the potential risks effectively (Ivanov 2021; AlMulhim 2021). Hence, it is essential for the organizations to prioritize the adoption and execution of smart technologies (Abbasnejad et al. 2020) to improve supply chain performance and profitability (Chege et al. 2020; AlMulhim 2021; Modgil and Sakhuja 2022; Manoram and Panghal 2023). Smart technologies refer to modern technological systems that automate operations by making use of novel technologies, i.e., blockchain technologies, AI technologies, machine learning, IoT, etc., (Chen 2022; Ali et al. 2023). The substantial benefits of advanced technologies are evident from the increasing potential of integrated applications and software components, as well as cross-platform, infrastructure, and manufacturing systems (Yoo 2010; Nasiri et al. 2020; AlMulhim 2021). However, the optimal utilization of digital transformation may require the implementation of smart technologies (Kallinikos et al. 2013; Chen 2022). As the utilization of smart technologies, such as IoT sensors and RFID tags, facilitates the real-time tracking and monitoring of supplies, which potentially enables transparency in supply chain operations (Unhelkar et al. 2022; Matarneh et al. 2024). Also, improved visibility and transparency across the supply chain can greatly enhance operational effectiveness (Swift et al. 2019). Furthermore, smart technologies provide a digital platform that connects the database side of integrated systems with the user interface of web-based systems across different partners to support an e-business initiative efficiently (Zhu et al. 2015). Consequently, according to Gecevska et al. (2012), in order to achieve a high degree of integration between processes is essential for achieving favorable business objectives. Therefore, the formation of connectivity among different parties and processes within the supply chain is essential for the betterment of supply chain performance (Anand and Ward 2009; Fatorachian and Kazemi 2020). Drawing from the aforementioned considerations, it can be concluded that the implementation of smart technologies has become essential for enhancing the digital transformation of organizations. Therefore, the following hypothesis can be proposed.

H3: *Smart technologies positively influence the supply chain performance of any organization.*

9.2.2.4 Role of Smart Technologies as Mediator Between DT and SCP

Complex and continuously shifting business conditions and consumer preferences, as well as the need for organizations to be agile and responsive, are putting emphasis on technological advancements (Lang et al. 2014; Öberg and Graham 2016; Büyüközkan and Göçer 2018; Chege et al. 2020; Verhoef et al. 2021). Therefore, most of the companies have recognized the importance of technological innovations and view technology as an effective strategic tool for achieving organizational performance (Chavarría-Barrientos et al. 2017). In doing so, numerous enterprises are embracing digital transformation by deploying e-business technologies to optimize their business operations globally (Chen et al. 2016; Chege et al. 2020; Abbasnejad et al. 2020). In addition to this, companies are making significant investments in smart technologies such as AI, machine learning, blockchain, IoT, robotics, and automation in order to embrace the most advanced technical developments (Fatorachian and Kazemi 2020; Manoram and Panghal 2023; Nazir et al. 2023) to enhance their supply chain efficiencies by attaining process integration (Liu et al. 2016; Schmidt et al. 2022). The exchange of information is known as an essential element in ensuring the management of connections within the supply chain and is considered as a key component of supply chain performance (Baihaqi and Sohal 2012; Abbasnejad et al. 2020). Therefore, businesses are investing in technological advancements to create efficient, transparent, and reliable communication channels to improve their supply chain performance (Govindan et al. 2017; Nazir et al. 2023). Considering the significant role of smart technologies in digitally transforming the firm's business practices and enhancing supply chain performance, the following hypothesis has been proposed.

> **H4**: *Smart technologies mediate the relationship between DT and supply chain performance.*

9.3 Conceptual Model

The deployment of smart technologies plays an essential role in digitally transforming the firm's various sectors, including the supply chain. Therefore, this study aims to investigate how the use of smart technologies can digitally transform the firm's businesses by improving their supply chain performance. As the success of a company's digital transformation depends on the unique qualities of the digital tools it chooses to implement (Nasiri et al. 2020; Fatorachian and Kazemi 2020; AlMulhim 2021; Manoram and Panghal 2023; Sucupira Furtado et al. 2023). In doing so, the utilization of smart and intelligent technologies is a crucial instrument in facilitating the conversion of a company's digital transformation into enhanced supply chain performance of an organization. Thus, the utilization of smart technologies serves as an acceleration for companies to improve their supply chain performance through

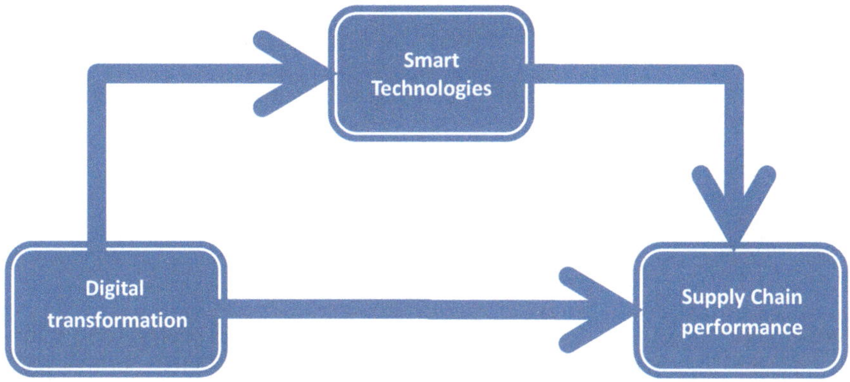

Fig. 9.2 Conceptual framework

digital transformation as the conceptual framework in (Fig. 9.2), representing the study's framework.

9.4 Research Methodology

9.4.1 Population and Sampling

The research study focuses on the local and multinational companies (MNCs) operating in Pakistan as they are the major contributor to the economy of the country. The presence of local manufacturing and multinational corporations in Pakistan has resulted in several positive effects on the economy such as infrastructure development, employment generation, introduction to innovative technologies, income tax revenue generation, local workforce training, and increased exports resulting in improved foreign exchange reserves. In 2020–21, the State Bank of Pakistan (SBP) reported a 2.8% increase in manufacturing employment, where most MNCs operate. Furthermore, according to Federal Board of Revenue (FBR) reported a 19.9% rise in MNCs tax income in 2020–21 (Saleem 2023). The data were collected from 41 local and MNCs working professionals from February 2023 to April 2023. The companies were contacted by personal visits as well as electronically through various means.

9.4.2 Data Collection Instrument

The data has been obtained via a closed-ended questionnaire. All the construct's items were taken from reliable scales that have been used previously. A total of four items of smart technologies were adopted from Yoo (2010) and Nasiri et al. (2020).

Table 9.1 Constructs

Construct	Items	Sources
Digital transformation	3	Berman (2012), Leviäkangas (2016), Frank et al. (2019), Pramanik et al. (2019)
Smart technologies	4	Yoo (2010), Nasiri et al. (2020)
Supply chain performance	3	Beamon (1999), Holmberg (2000), Sillanpää (2015)

Furthermore, three items were used to assess the companies' capacity to utilize digital technology in their operations as a means of measuring digital transformation adopted from Berman (2012), Leviäkangas (2016), Frank et al. (2019), Pramanik et al. (2019). Likewise, three items of construct supply chain performance were adopted from Beamon (1999), Holmberg (2000), and Sillanpää (2015). The independent variable digital transformation, smart technologies (mediating) and dependent variable (supply chain performance) were measured on a scale from 1 to 5, i.e., strongly disagree (1), disagree (2), neither agree nor disagree (3), agree (4), and strongly agree (5) accordingly. Table 9.1 exhibits the constructs, items and their sources. This research comprises of local and multinational companies and their acceptance level of technology and its impact on their supply chain performance (SCP).

9.4.3 Analytical Methods

In this research, PLS-SEM has been used instead of other methods because it is more stable and robust. PLS-SEM is performed in two parts (Dias et al. 2021; Khan et al. 2022). In the initial phase, the reliability, validity, and model fitness are assessed. While at the second phase, path analysis is performed for hypotheses tastings. In addition, multi-group analysis with PLS has been performed to examine the results.

9.5 Results

9.5.1 Reliability, Validity, and Fitness of Constructs

This research study followed the criteria suggested by Hair et al. (2016) to assess the validity and reliability of the constructs. Firstly, it assessed CR and CB alpha values to determine construct reliability. According to Hair et al. (2016), CB alpha and CR values are 0.70 required to determine the constructs internal reliability. The findings are presented in Table 9.2, which demonstrates that each of the construct's values is more than 0.70 except the values of digital transformation of the local group which is 0.639. However, according to Hair et al. (2006), values close to 0.60 are acceptable, particularly if the factor has few items. In addition, this table also exhibits the factor loading values to ensure reliability. All the items have factor loadings that are 0.815 or greater, indicating the stable indicator's reliability. In addition, the AVE findings are also presented in Table 9.2, which exhibits that all constructs values of AVE are greater than 0.50. According to Hair et al. (2016), for a construct to be considered convergently valid, its AVE values must be higher than 0.50. Thus, all constructs possess convergent validity. Likewise, Table 9.3 exhibits the Fornell-Larcker criteria that have been used to determine the discriminant validity of our tests. The diagonal values, square-rooted AVE values, are higher than inter-construct correlation, confirming all constructs' discriminant validity. Collectively, the analyses of AVE, CR, loadings, CB alpha, and fitness demonstrate the constructs' reliability, validity, and fitness (Tables 9.4 and 9.5).

Table 9.2 Reliability, validity, and model fitness

Constructs	Loadings	Local group	MNC's group	Local group	MNC's group	Local group	MNC's group
		Cronbach's alpha	Cronbach's alpha	Composite reliability	Composite reliability	Average variance extracted (AVE)	Average variance extracted (AVE)
Digital transformation	0.828	0.639	0.758	0.804	0.776	0.550	0.671
	0.862						
	0.835						
Smart technologies	0.844	0.742	0.839	0.767	0.929	0.559	0.673
	0.856						
	0.841						
	0.855						
Supply chain performance	0.815	0.742	0.705	0.766	0.709	0.658	0.630
	0.820						
	0.888						

Table 9.3 Fornell-Larcker criteria

	Digital transformation	Smart technologies	Supply chain performance
Digital transformation	0.842		
Smart technologies	0.717	0.849	
Supply chain performance	0.696	0.636	0.842

Table 9.4 Variance inflation factor

	Items	Total	Local group	MNC's group
		VIF	VIF	VIF
Digital transformation	DT1	1.812	1.300	1.842
	DT2	1.555	1.177	1.444
	DT3	1.836	1.370	1.546
Smart technologies	SCP1	1.632	1.424	1.257
	SCP2	1.631	1.452	1.510
	SCP3	1.887	1.629	1.471
	ST1	2.849	2.150	1.373
Supply chain performance	ST2	3.077	2.386	2.232
	ST3	2.566	1.852	2.684
	ST4	2.775	2.034	3.311

Table 9.5 Structural model hypotheses testing (individual group)

	Total	Local group	MNC's group	Total	Local group	MNC's group
	Beta	Beta	Beta	*P* values	*P* values	*P* values
Digital transformation → smart technologies	0.719	0.640	0.434	0.000	0.000	0.000
Digital transformation → supply chain performance	0.496	0.354	0.695	0.000	0.052	0.000
Smart technologies → supply chain performance	0.285	0.216	0.291	0.001	0.256	0.000
Digital transformation → smart technologies → supply chain performance	0.206	0.120	0.125	0.002	0.283	0.003

9.5.2 Path Analysis

The results of the performed hypotheses tests are presented in Table 9.6. The findings from the first test of hypotheses are ($\beta = 0.719$, t value 15.523) demonstrate a significant direct impact of digital transformation on smart technologies, thus supporting the H1. Furthermore, the findings indicate that ($\beta = 0.496$, t value 5.342) there is a significant direct effect of digital transformation on firm's SCP also supporting the H2. Likewise, the findings also demonstrate that smart technologies have a substantial effect on a firm's SC performance (p-value = 0.285, t-value = 3.257), thus supporting the H3. Moreover, these results confirm that smart technologies play a significant mediating role between digital transformation and firms SC performance which is (p-value = 0.206, t value = 3.058), also supporting the H4.

In order to assess the consistency of the findings regarding the technology adoption by a firm, the dataset has been partitioned based on local and multinational companies (MNCs) and performed the multi-group analysis (PLS-MGA), the results are presented in Table 9.7. The findings present a substantial difference between both groups. Furthermore, Harman's one-factor test (Harman 1976) has been used to detect common method biases. The results of the Harman test are supported by the observations that the VIF values are lower than 3.

Table 9.6 Structural model hypotheses testing

Relationship(s)	Beta	S.E.	t value	Decision	
				p value	Accept/reject
Digital transformation → smart technologies	0.719	0.046	15.523	0.000	Accepted
Digital transformation → supply chain performance	0.496	0.092	5.342	0.000	Accepted
Smart technologies → supply chain performance	0.285	0.087	3.257	0.001	Accepted
Digital transformation → smart technologies → supply chain performance	0.206	0.066	3.058	0.002	Accepted

Table 9.7 Multi-group analysis

Hypotheses		Local group	MNC's group	Difference	(P) p value
H1	Digital transformation → smart technologies	0.795	0.484	0.311	0.000
H2	Digital transformation → supply chain performance	0.598	1.100	– 0.503	0.000
H3	Smart technologies → supply chain performance	0.165	0.231	– 0.066	0.000
H4	Digital transformation → smart technologies → supply chain performance	0.131	0.112	0.019	0.000

9.6 Discussion

This study examined the role of smart technologies as a mediator between digital transformation and firms SC performance. The findings indicate that various supply chain functions, including inventory management, processing of orders, delivery, and shipment traceability could be automated through digital transformation. The incorporation of technological innovations into every aspect of a business significantly alters the way companies function and provide value to customers, resulting in a new era of transformation. This automation may have the potential to assist organizations to eliminate errors and boost efficiency while enhancing overall supply chain performance. Furthermore, the study suggests that technological investments by a firm may lead to produce significant benefits for organizations. Such as the incorporation of smart technologies within the supply chain possesses the capability to transform supply chain activities, i.e., increasing efficiency, lowering costs, and improving customer satisfaction. Thus, the utilization of smart technologies such as blockchain, artificial intelligence, and IoT has the potential to increase supply chain visibility, reduce lead time and promote better decision-making capabilities (AlMulhim 2021; Yu et al. 2022; Nazir et al. 2023; Ozmen and Bora 2023; Ali et al. 2023).

Likewise, as a second contribution, the findings of this study also suggests that digital transformation profoundly impacts supply chain performance, enabling firms to enhance efficiency, responsiveness, and overall competitiveness. Furthermore, it allows them to collect and analyze data on real-time basis (Gezgin et al. 2017). These technological shifts allow organizations to enhance agility, lower costs, and streamline their business operations including supply chain. Consequently, these developments not only lower costs and make the supply chain more flexible, but they also make customers happier and make sure that rules are followed, which makes the supply chain more competitive and successful. Moreover, this research's contributions also support the previous study conducted by (Rehman Khan et al. 2022).

Furthermore, the findings also demonstrate that smart technologies offer an actual means to execute digital transformation within an organization. However, the adoption of smart technologies poses a challenge and requires a considerable amount of time. Therefore, companies must possess specific skills and comprehend their objectives to effectively implement such technologies efficiently. Furthermore, it is essential for companies to extensively choose suitable digital technologies that are in line with the organization's strategic goals and effectively incorporate them into their business operations. By doing so, companies can achieve better performance outcomes. The results of this study also indicate that the adoption of smart technologies enhances the ability of product tracking in real time, leading to a considerable reduction in the total expenses incurred during the transportation of goods across the supply chain. Overall, the implementation of digital transformation can assist companies in utilizing smart technology to enhance their supply chain performance (Fig. 9.3).

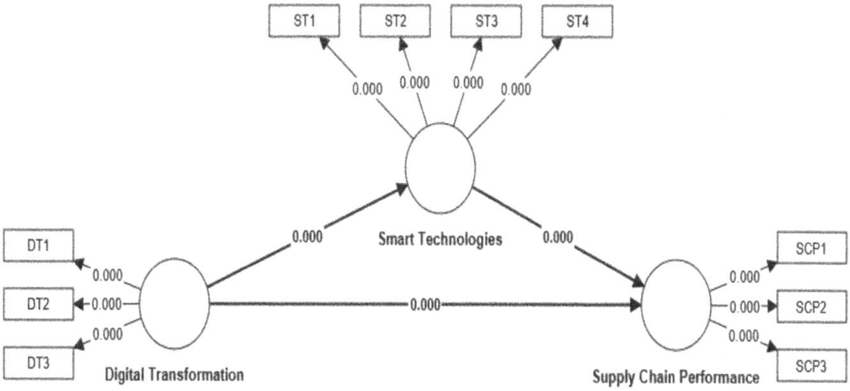

Fig. 9.3 Measurement model

9.7 Conclusion, Managerial Implications, Limitations, and Future Research

9.7.1 Conclusion

The study's main objective was to investigate the significant role of digital transformation on the firms SC performance through the adoption of smart technologies. The study conducted an analysis of 120 individuals working in 41 local and multinational firms in Pakistan, utilizing PLS-SEM and PLS-Multi-Group-Analysis. The results indicated a noteworthy direct impact of digital transformation on the firms SC performance. Likewise, the results also present the positive effect of smart technologies on firms SC performance. Furthermore, the results also demonstrate the positive mediating effect of smart technologies in the connection between digital transformation and firms SC performance.

9.7.2 Academic Implications

Furthermore, this research study made a substantial contribution to the existing body of literature on the subject of digital transformation, smart technologies, and the supply chain performance of manufacturing industries. The findings of this research study suggest that the incorporation of smart technologies is essential for the digital transformation of firms and enhance their SC performance. The research results also indicate that the implementation of smart technologies enhances the firm's ability to reduce lead time, and track products in real-time leading to a substantial decrease in overall costs associated with the movement of goods within a supply chain.

9.7.3 Managerial Implications

Finally, our present study suggests that the governing bodies should incorporate this development into their existing framework in order to facilitate the digitization of manufacturing industries. Likewise, the aforementioned strategy enables organizations to potentially integrate digitalization into their business operations including supply chain, allocate resources toward smart technology, and capitalize on the digitalization trend to enhance firms supply chain performance. Hence, it is essential for managers to focus on the implementation of digital transformation strategies, with a specific emphasis on enhancing supply chain performance and maintaining long-term stability within the global market.

9.7.4 Limitations and Future Research

Similar to other studies in this field, this research study exhibits certain limitations. One primary constraint of this research is the inability of the utilization of longitudinal data due to time constraints. Nevertheless, the utilization of longitudinal data has the potential to uncover the authentic essence of smart technologies on a firm's performance. But the applicability of smart technologies is time taking process and requires substantial initial investment, they are not readily applicable. Therefore, in this research, the data has been collected from two distinct groups, i.e., (national and multinational companies) in order to determine the feasibility of implementing smart technologies. The other constraint of this study is its narrow scope, as it solely concentrates on developing countries, and therefore the application of findings in any other country context should be adopted with caution. Similarly, the organizations' infrastructure and business operations are well-structured in developed countries. Therefore, the findings could be very different if this research study were carried out in developed countries, and conducting a cross-country analysis could offer an extensive result.

Acknowledgements During the preparation of this work, the author(s) used AI tools to create and proofread some parts of the manuscript. After using this tool/service, the author(s) reviewed and edited the content as needed and take(s) full responsibility for the content of the publication.

References

Abbasnejad, B., Nepal, M. P., Ahankoob, A., Nasirian, A., & Drogemuller, R. (2020). Building Information Modelling (BIM) adoption and implementation enablers in AEC firms: A Systematic Literature Review. *Architectural Engineering and Design Management*, *17*(5–6), 411–433. https://doi.org/10.1080/17452007.2020.1793721.

Agrawal, P., & Narain, R. (2018). Digital Supply Chain Management: An overview. *IOP Conference Series: Materials Science and Engineering, 455*, 012074. https://doi.org/10.1088/1757-899x/455/1/012074.

Ali, M., Nazir, S., & Junaid, M. (2023). Blockchain Driven Supply Chain Management and Supply Chain Resilience: Role of Intellectual Capital. *Management for Professionals*, 239–254. https://doi.org/10.1007/978-981-99-0699-4_14.

AlMulhim, A. F. (2021). Smart Supply Chain and firm performance: The Role of Digital Technologies. *Business Process Management Journal, 27*(5), 1353–1372. https://doi.org/10.1108/bpmj-12-2020-0573.

Anand, G., & Ward, P. T. (2009). Fit, flexibility and performance in manufacturing: Coping with Dynamic Environments. *Production and Operations Management, 13*(4), 369–385. https://doi.org/10.1111/j.1937-5956.2004.tb00224.x.

Attaran, M. (2012). Critical Success Factors and Challenges of Implementing RFID in Supply Chain Management. *Journal of Supply Chain and Operations Management, 10*(1), 144–167.

Attaran, M. (2020). Digital technology enablers and their implications for Supply Chain Management. *Supply Chain Forum: An International Journal, 21*(3), 158–172. https://doi.org/10.1080/16258312.2020.1751568.

Baihaqi, I., & Sohal, A. S. (2012). The impact of information sharing in supply chains on organisational performance: An empirical study. *Production Planning & Control, 24*(8–9), 743–758. https://doi.org/10.1080/09537287.2012.666865.

Beamon, B. M. (1999). Measuring Supply Chain Performance. *International Journal of Operations & Production Management, 19*(3), 275–292. https://doi.org/10.1108/01443579910249714.

Berman, S. J. (2012). Digital Transformation: Opportunities to create new business models. *Strategy & Leadership, 40*(2), 16–24. https://doi.org/10.1108/10878571211209314.

Büyüközkan, G., & Göçer, F. (2018). Digital Supply Chain: Literature Review and A proposed framework for future research. *Computers in Industry, 97*, 157–177. https://doi.org/10.1016/j.compind.2018.02.010.

Chavarría-Barrientos, D., Batres, R., Wright, P. K., & Molina, A. (2017). A methodology to create a sensing, smart and Sustainable Manufacturing Enterprise. *International Journal of Production Research, 56*(1–2), 584–603. https://doi.org/10.1080/00207543.2017.1386333.

Chege, S. M., Wang, D., & Suntu, S. L. (2020). Influence of technology innovation intensity on firm performance. *International Journal of Technology and Human Interaction, 16*(2), 34–52. https://doi.org/10.4018/ijthi.2020040104.

Chen, X. (2022). Smart Technologies and Aging Society. *Smart Innovation, Systems and Technologies*, 131–146. https://doi.org/10.1007/978-981-19-1146-0_7.

Chen, X., Despeisse, M., & Johansson, B. (2020). Environmental sustainability of digitalization in manufacturing: A Review. *Sustainability, 12*(24), 10298. https://doi.org/10.3390/su122410298.

Chen, Y.-Y. K., Jaw, Y.-L., & Wu, B.-L. (2016). Effect of digital transformation on organisational performance of smes. *Internet Research, 26*(1), 186–212. https://doi.org/10.1108/intr-12-2013-0265.

Chopra, S. (2020). *Supply Chain Management: Strategy, planning and operation*. Pearson Education.

Dias, C., Gouveia Rodrigues, R., & Ferreira, J. J. (2021). Small agricultural businesses' performance—what is the role of dynamic capabilities, entrepreneurial orientation, and environmental sustainability commitment? *Business Strategy and the Environment, 30*(4), 1898–1912. https://doi.org/10.1002/bse.2723.

Dreyer, H. C., Strandhagen, J. O., Hvolby, H.-H., Romsdal, A., & Alfnes, E. (2016). Supply Chain Strategies for speciality foods: A Norwegian case study. *Production Planning & Control, 27*(11), 878–893. https://doi.org/10.1080/09537287.2016.1156779.

Elsersy, M., Sherif, A., Darwsih, A., & Hassanien, A. E. (2021). Digital Transformation and emerging technologies for tackling COVID-19 pandemic. *Digital Transformation and Emerging*

Technologies for Fighting COVID-19 Pandemic: Innovative Approaches, 3–19. https://doi.org/10.1007/978-3-030-63307-3_1.

Fabbe-Costes, N., & Lechaptois, L. (2022). Automotive Supply Chain Digitalization. *The Digital Supply Chain,* 289–308. https://doi.org/10.1016/b978-0-323-91614-1.00017-4.

Fatorachian, H., & Kazemi, H. (2020). Impact of industry 4.0 on Supply Chain Performance. *Production Planning & Control, 32*(1), 63–81. https://doi.org/10.1080/09537287.2020.1712487.

Frank, A. G., Dalenogare, L. S., & Ayala, N. F. (2019). Industry 4.0 technologies: Implementation patterns in manufacturing companies. *International Journal of Production Economics, 210,* 15–26. https://doi.org/10.1016/j.ijpe.2019.01.004.

Gecevska, V., Veza, I., Cus, F., Anisic, Z. and Stefanic, N., 2012. Lean PLM-information technology strategy for innovative and sustainable business environment. International Journal of Industrial Engineering and Management, 3(1), pp. 15–23.

Gezgin, E., Huang, X., Samal, P. and Silva, I., 2017. Digital transformation: Raising supply-chain performance to new levels. *McKinsey & Company*, pp. 1–10.

Govindan, K., Mangla, S. K., & Luthra, S. (2017). Prioritising indicators in improving supply chain performance using Fuzzy AHP: Insights from the case example of four Indian manufacturing companies. *Production Planning & Control, 28*(6–8), 552–573. https://doi.org/10.1080/09537287.2017.1309716.

Gupta, H., Yadav, A. K., Kusi-Sarpong, S., Khan, S. A., & Sharma, S. C. (2022). Strategies to overcome barriers to innovative digitalisation technologies for Supply Chain Logistics Resilience during pandemic. *Technology in Society, 69,* 101970. https://doi.org/10.1016/j.techsoc.2022.101970.

Hair, J. F., Black, W. C., Babin, B. J., Anderson, R. E., & Tatham, R. L. (2006). *Multivariate Data Analysis.* Upper Saddle River, NJ: Pearson Prentice Hall.

Hair, Jr., J. F., Sarstedt, M., Matthews, L. M., & Ringle, C. M. (2016). Identifying and treating unobserved heterogeneity with FIMIX-PLS: Part I – method. *European Business Review, 28*(1), 63–76. https://doi.org/10.1108/ebr-09-2015-0094.

Harman, H. H. (1976). *Modern Factor Analysis.* Chicago: University of Chicago Press.

Holmberg, S. (2000). A systems perspective on supply chain measurements. *International Journal of Physical Distribution & Logistics Management, 30*(10), 847–868. https://doi.org/10.1108/09600030010351246.

Ivanov, D. (2021). Supply Chain Risks, disruptions, and ripple effect. *Introduction to Supply Chain Resilience,* 1–28. https://doi.org/10.1007/978-3-030-70490-2_1.

Ivanov, D., Dolgui, A., & Sokolov, B. (2018). The impact of digital technology and Industry 4.0 on the ripple effect and supply chain risk analytics. *International Journal of Production Research, 57*(3), 829–846. https://doi.org/10.1080/00207543.2018.1488086.

Ivanov, D. and Sokolov, B. (2010) 'Evolution of supply chain management (SCM)', *Adaptive Supply Chain Management*, pp. 1–17. https://doi.org/10.1007/978-1-84882-952-7_1.

Ji, G., Zhou, S., Lai, K., Tan, K. H., & Kumar, A. (2022). Timing of blockchain adoption in a supply chain with competing manufacturers. *International Journal of Production Economics, 247,* 108430. https://doi.org/10.1016/j.ijpe.2022.108430.

Junaid, M., Nazir, S., & Ali, M. (2023). Role of Blockchain Technology Adoption between Sustainability Related Supply Chain Risks and Triple Bottom Line Performance. In *Blockchain Driven Supply Chain Management: A Multi-dimensional Perspective* (pp. 181–199). Singapore: Springer Nature Singapore.

Kallinikos, J., Aaltonen, A., & Marton, A. (2013). The ambivalent ontology of digital artifacts. *MIS Quarterly, 37*(2), 357–370. https://doi.org/10.25300/misq/2013/37.2.02.

Kearney. (2015, September 9). Digital supply chains: Increasingly critical for competitive edge. Retrieved April 20, 2023, from https://www.kearney.com/operations-performance-transformation/article/-/insights/digital-supply-chains-increasingly-critical-for-competitive-edge.

Khan, S. A., Mubarik, M. S., Kusi-Sarpong, S., Gupta, H., Zaman, S. I., & Mubarik, M. (2022). Blockchain technologies as enablers of supply chain mapping for Sustainable Supply Chains. *Business Strategy and the Environment, 31*(8), 3742–3756. https://doi.org/10.1002/bse.3029.

Lang, M., Deflorin, P., Dietl, H., & Lucas, E. (2014). The impact of complexity on knowledge transfer in manufacturing networks. *Production and Operations Management, 23*(11), 1886–1898. https://doi.org/10.1111/poms.12193.

Leviäkangas, P. (2016). Digitalisation of Finland's transport sector. *Technology in Society, 47*, 1–15. https://doi.org/10.1016/j.techsoc.2016.07.001.

Li, F. (2020). The digital transformation of business models in the Creative Industries: A Holistic Framework and Emerging Trends. *Technovation, 92–93*, 102012. https://doi.org/10.1016/j.technovation.2017.12.004.

Li, H., & Zhang, Q. (2023). Research on Digital Transformation of Supply Chain: Research Status and recommendations. *SHS Web of Conferences, 153*, 01001. https://doi.org/10.1051/shsconf/202315301001.

Liu, Z., Prajogo, D., & Oke, A. (2016). Supply Chain Technologies: Linking adoption, utilization, and performance. *Journal of Supply Chain Management, 52*(4), 22–41. https://doi.org/10.1111/jscm.12117.

Manoram, S., & Panghal, A. (2023). Smart Technologies Interventions for Sustainable Agri-Food Supply Chain. *Digital Transformation and Industry 4.0 for Sustainable Supply Chain Performance,* 79–100. https://doi.org/10.1007/978-3-031-19711-6_3.

Matarneh, S. *et al.* (2024) 'Industry 4.0 technologies and circular economy synergies: Enhancing corporate sustainability through sustainable supply chain integration and flexibility', *Environmental Technology & Innovation*, 35, p. 103723. https://doi.org/10.1016/j.eti.2024.103723.

Michel, R. (2017, May 05). The evolution of the Digital Supply Chain. Retrieved April 10, 2023, from https://www.supplychain247.com/article/the_evolution_of_the_digital_supply_chain.

Modgil, A., & Sakhuja, S. (2022). The role of Smart Technologies in enhancement of supply chain visibility. *ECS Transactions, 107*(1), 8113–8120. https://doi.org/10.1149/10701.8113ecst.

Mokhtar, S., Bahri, P. A., Moayer, S., & James, A. (2017). A novel decision-making approach for supplier selection under risks. *Computer Aided Chemical Engineering,* 1267–1272. https://doi.org/10.1016/b978-0-444-63965-3.50213-0.

Najmi, A., & Makui, A. (2011). A conceptual model for measuring supply chain's performance. *Production Planning & Control, 23*(9), 694–706. https://doi.org/10.1080/09537287.2011.586004.

Nasiri, M., Ukko, J., Saunila, M., & Rantala, T. (2020). Managing the Digital Supply Chain: The role of smart technologies. *Technovation, 96–97*, 102121. https://doi.org/10.1016/j.technovation.2020.102121.

Nazir, S., Junaid, M., Ali, M., & Saad, M. (2023). Role of intellectual capital in implementing blockchain technology-driven sustainable supply chain: A proposed framework. *Management for Professionals,* 201–218. https://doi.org/10.1007/978-981-99-0699-4_12.

Nazir, S. *et al.* (2024) 'Sustainable performance and disaster management in the oil and gas industry: An Intellectual Capital Perspective', *Resources Policy*, 92, p. 105042. https://doi.org/10.1016/j.resourpol.2024.105042.

Öberg, C., & Graham, G. (2016). How smart cities will change supply chain management: A technical viewpoint. *Production Planning & Control, 27*(6), 529–538. https://doi.org/10.1080/09537287.2016.1147095.

Omar, I. A., Debe, M., Jayaraman, R., Salah, K., Omar, M., & Arshad, J. (2022). Blockchain-based supply chain traceability for covid-19 personal protective equipment. *Computers & Industrial Engineering, 167*, 107995. https://doi.org/10.1016/j.cie.2022.107995

Ozmen, E., & Bora, N. (2023). Blockchain, IOT, and Artificial Intelligence Technologies for Supply Chain Management: Bibliometric Analysis. *Blockchain, IoT, and AI Technologies for Supply Chain Management,* 19–41. https://doi.org/10.1201/9781003264521-2.

Pramanik, H. S., Kirtania, M., & Pani, A. K. (2019). Essence of digital transformation—manifestations at large financial institutions from North America. *Future Generation Computer Systems, 95*, 323–343. https://doi.org/10.1016/j.future.2018.12.003.

Raj, P. V., Jauhar, S. K., Ramkumar, M., & Pratap, S. (2022). Procurement, traceability and advance cash credit payment transactions in supply chain using blockchain smart contracts. *Computers & Industrial Engineering, 167*, 108038. https://doi.org/10.1016/j.cie.2022.108038.

Rehman Khan, S. A., Ahmad, Z., Sheikh, A. A., & Yu, Z. (2022). Digital Transformation, Smart Technologies, and eco-innovation are paving the way toward sustainable supply chain performance. *Science Progress, 105*(4), 003685042211456. https://doi.org/10.1177/00368504221145648.

Rapaccini, M., Saccani, N., Kowalkowski, C., Paiola, M., & Adrodegari, F. (2020). Navigating disruptive crises through service-led growth: The impact of covid-19 on Italian manufacturing firms. *Industrial Marketing Management, 88*, 225–237. https://doi.org/10.1016/j.indmarman.2020.05.017.

Saberi, S., Kouhizadeh, M., Sarkis, J., & Shen, L. (2018). Blockchain technology and its relationships to sustainable supply chain management. *International Journal of Production Research, 57*(7), 2117–2135. https://doi.org/10.1080/00207543.2018.1533261.

Saleem, F. (2023, March 29). Impact of multinational companies on Pakistan. Retrieved April 28, 2023, from https://moderndiplomacy.eu/2023/03/30/impact-of-multinational-companies-on-pakistan/.

Schmidt, I., Morris, W., Thomas, A., & Manning, L. (2022). Smart systems: The role of advanced technologies in improving business quality, performance and Supply Chain Integration. *Standards, 2*(3), 276–293. https://doi.org/10.3390/standards2030020.

Schroeck, M., Kwan, A., Kawamura, J., Stefanita, C., & Sharma, D. (2019, June 16). Digital Industrial Transformation. Retrieved April 30, 2023, from https://www.deloitte.com/global/en/our-thinking/insights/topics/digital-transformation/industry-4-0/digital-industrial-transformation-industrial-internet-of-things.html.

Sherman, E. (2020). 94% of the fortune 1000 are seeing coronavirus supply chain disruptions. Retrieved from https://fortune.com/2020/02/21/fortune-1000-coronavirus-china-supply-chain-impact/.

Shokouhifar, M., & Ranjbarimesan, M. (2022). Multivariate time-series blood donation/demand forecasting for resilient supply chain management during COVID-19 pandemic. *Cleaner Logistics and Supply Chain, 5*, 100078. https://doi.org/10.1016/j.clscn.2022.100078.

Sillanpää, I. (2015). Empirical study of measuring supply chain performance. *Benchmarking: An International Journal, 22*(2), 290–308. https://doi.org/10.1108/bij-01-2013-0009.

Sucupira Furtado, L., da Silva, T. L., Ferreira, M. G., de Macedo, J. A., & de Melo Lima Cavalcanti Moreira, J. K. (2023). A framework for Digital Transformation towards smart governance: Using big data tools to target sdgs in Ceará, Brazil. *Journal of Urban Management, 12*(1), 74–87. https://doi.org/10.1016/j.jum.2023.01.003.

Swift, C., Guide, V. D., & Muthulingam, S. (2019). Does supply chain visibility affect operating performance? evidence from conflict minerals disclosures. *Journal of Operations Management, 65*(5), 406–429. https://doi.org/10.1002/joom.1021.

Tietze, F., Vimalnath, P., Aristodemou, L., & Molloy, J. (2022). Crisis-critical intellectual property: Findings from the COVID-19 pandemic. *IEEE Transactions on Engineering Management, 69*(5), 2039–2056. https://doi.org/10.1109/tem.2020.2996982.

Unhelkar, B., Joshi, S., Sharma, M., Prakash, S., Mani, A. K., & Prasad, M. (2022). Enhancing supply chain performance using RFID technology and decision support systems in the industry 4.0—A systematic literature review. *International Journal of Information Management Data Insights, 2*(2), 100084. https://doi.org/10.1016/j.jjimei.2022.100084.

Vanteddu, G., & Bandyopadhayay, A. (2022). Blockchain technology aptness for improving supply chain visibility, resiliency, and efficiency. *Handbook of Research on Supply Chain Resiliency, Efficiency, and Visibility in the Post-Pandemic Era,* 316–334. https://doi.org/10.4018/978-1-7998-9506-0.ch016.

Verhoef, P. C., Broekhuizen, T., Bart, Y., Bhattacharya, A., Qi Dong, J., Fabian, N., & Haenlein, M. (2021). Digital Transformation: A multidisciplinary reflection and research agenda. *Journal of Business Research, 122*, 889–901. https://doi.org/10.1016/j.jbusres.2019.09.022.

Warner, K. S. R., & Wäger, M. (2019). Building dynamic capabilities for digital transformation: An ongoing process of strategic renewal. *Long Range Planning, 52*(3), 326–349. https://doi.org/10.1016/j.lrp.2018.12.001.

Xu, Z., Elomri, A., Kerbache, L., & El Omri, A. (2020). Impacts of covid-19 on Global Supply Chains: Facts and perspectives. *IEEE Engineering Management Review, 48*(3), 153–166. https://doi.org/10.1109/emr.2020.3018420.

Ye, F., Liu, K., Li, L., Lai, K.-H., Zhan, Y., & Kumar, A. (2022). Digital supply chain management in the COVID-19 crisis: An Asset Orchestration Perspective. *International Journal of Production Economics, 245*, 108396. https://doi.org/10.1016/j.ijpe.2021.108396.

Yoo. (2010). Computing in everyday life: A call for research on experiential computing. *MIS Quarterly, 34*(2), 213. https://doi.org/10.2307/20721425.

Yu, Z., Shah, A., Rehman, S. A., Nazir, S., & Tanveer, M. (2022). Blockchain technology and sustainable supply chain practices: Leading towards organizational performance. *Journal of Advanced Manufacturing Systems*. https://doi.org/10.1142/s0219686723500257.

Zahid, R. M., Khurshid, M., & Ying, C. (2023). Digital Technologies and Supply Chain Management. *Emerging Trends in Sustainable Supply Chain Management and Green Logistics,* 41–74. https://doi.org/10.4018/978-1-6684-6663-6.ch003.

Zhong, R. Y., Xu, X., & Wang, L. (2017). IOT-enabled smart factory visibility and traceability using laser-scanners. *Procedia Manufacturing, 10*, 1–14. https://doi.org/10.1016/j.promfg.2017.07.103.

Zhu, H. (2022). Digital Transformation of Smart Industries and transformation of smart city governance models in the context of internet technology and Digital Economy. *Proceedings of the 1st International Conference on Public Management, Digital Economy and Internet Technology*. https://doi.org/10.5220/0011750800003607.

Zhu, Z., Zhao, J., Tang, X., & Zhang, Y. (2015). Leveraging e-business process for business value: A layered structure perspective. *Information & Management, 52*(6), 679–691. https://doi.org/10.1016/j.im.2015.05.004.

Chapter 10
Leveraging Open Innovation for Smart Supply Chain

Sobia Jamil and Nabiha Kanwal

10.1 Introduction

The role of open innovation is increasingly prominent in supply chain management, offering new avenues for collaboration and improvement across various sectors, including higher education, the food industry, social enterprises, startups, small and medium-sized businesses, and the pharmaceutical industry. Open innovation involves leveraging both internal and external information, sources, and resources to create new products and services (Naqshbandi and Kamel 2017). This approach can significantly enhance the adoption and performance of smart supply chains by generating profits, increasing sales, enhancing brand recognition, achieving cost savings, improving productivity, and fostering capabilities to enter new markets, target different customer segments, and forge new external connections (Najib et al. 2021).

Smart supply chain management (SSCM) integrates technological advancements in supply chain activities, such as the flow of goods, information, and financials, enhancing transparency, resilience, sustainability, and adaptability (Qu and Kim 2024). Open innovation supports the adoption and implementation of SSCM by facilitating internal learning, technological innovation, and governance. Essential capabilities that support open innovation include internal learning, technological innovation, and governance. The concept of open innovation competency is proposed as a critical prerequisite for facilitating open innovation in supply chains. These skills enable companies to strategically manage knowledge inflows and outflows, thus enhancing internal innovation (Brunswicker and Vanhaverbeke 2015).

Effective knowledge creation within a supply chain is crucial for integrating and coordinating the diverse knowledge resources of all parties involved, boosting competitiveness (Jalees et al. 2021). The contractual mechanism is essential for

S. Jamil (✉) · N. Kanwal
Jinnah University for Women, Karachi, Pakistan
e-mail: sobiajamil901@gmail.com

M. S. Mubarik and S. A. Khan (eds.), *Smart Supply Chain Management*,
https://doi.org/10.1007/978-981-96-1333-5_10

formally establishing duties and providing trade partners with adequate control tools, understanding the prospective impact of legal and relational governance processes on effectively managing resource exchange in supply chains (Chiu and Lin 2022). Despite the benefits, controversies in the supply chain, such as the horse meat scandal, highlight the ongoing issues of transparency and traceability of raw materials. The COVID-19 pandemic further exposed vulnerabilities, with over 60% of businesses unaware of the disease's spread outside their local suppliers. There is a noticeable gap in the literature on combining distinctive export skills with cutting-edge production technology to meet diverse and unexpected customer requirements, despite numerous studies on open innovation and smart manufacturing systems (Lepore et al. 2023).

The implementation of open innovation in SSCM can significantly enhance the resilience and response of supply chains by integrating physical and conceptual resources, improving efficiency and adaptability in critical supply chain functions such as sourcing, production, distribution, and marketing. Newly emergent technologies like crowdsourcing, open innovation, industry 4.0, cloud manufacturing, IoT, and big data are shifting the Operations and Supply Chain Management (OSCM) paradigm, offering new opportunities for innovation and efficiency (Rahmanzadeh et al. 2022). For instance, smart contracts, a type of programmable computer language stored on blockchain systems, can streamline transactions and reduce the possibility of fraud or tampering, enhancing supply chain efficiency (Masa'deh et al. 2024). However, there is a lack of information in the literature on effectively fostering, establishing, and maintaining partnerships among various stakeholders within the complex framework of supply networks (Veldhuizen et al. 2013). Building mechanisms to source, share, filter, and evaluate external knowledge is essential (Enkel et al. 2020). This chapter aims to fill these gaps by conducting an integrative review focused on the connections between supply chain management and open innovation, constructing a framework to support the management of knowledge sources, and maximizing the innovative capacity embedded within organizations (Assis de Souza et al. 2023). This chapter contributes to the body of knowledge by highlighting the intersection of open innovation and smart supply chain management. It provides insights into how organizations can leverage external knowledge and digital advancements to enhance their supply chain operations and achieve a competitive edge.

The structure of this chapter is as follows: Sect. 10.2 discusses the motivation for this study. Section 10.3 provides an overview of SSCs. Section 10.4 offers an overview of AI. Section 10.5 explores the relationship between AI and SSCs. Section 10.6 examines the enablers and barriers to AI implementation in SSCs. Section 10.7 presents discussions and implications. Section 8 concludes the chapter. By addressing these areas, the chapter offers strategic guidance for researchers and practitioners aiming to harness the full potential of digitalization and open innovation in smart supply chain management.

10.2 Open Innovation and Smart Supply Chain: Background

Open Innovation and smart supply chains are interconnected concepts that strive to improve collaboration, efficiency, and competitiveness in the business sector. Open Innovation represents a paradigm shift in how organizations approach innovation. Traditionally, companies depended on internal research and development (R&D) for generating new ideas, technologies, and products. However, open innovation acknowledges that valuable knowledge and ideas can also originate externally. It highlights the significance of collaborating and sharing knowledge with external entities such as customers, suppliers, universities, startups, and even competitors (Papa et al. 2021). Key features of open innovation include: First, external collaboration encourages organizations to actively seek out external partners for innovation projects. This can include partnerships, joint ventures, licensing agreements, or even crowdsourcing ideas from the public. Second, Knowledge integration involves combining external knowledge with internal capabilities to create value. It recognizes that no single organization possesses all the necessary expertise and that external ideas and technologies can complement internal capabilities (de Araújo Burcharth et al. 2015). Third, value creation aims to leverage external knowledge to create value, reduce time to market, access new markets, and promote a culture of innovation within the organization. Fourth, Intellectual property management becomes more complex with open innovation, as organizations must balance protecting their intellectual property (IP) while sharing knowledge with external partners (Chesbrough and Vanhaverbeke 2018).

Supply chain management involves the coordination and integration of various activities related to the procurement, production, and delivery of goods and services to customers. Smart supply chains utilize digital technologies, data analytics, and automation to optimize operations, enhance visibility, and improve decision-making. Key features of smart supply chains include data-driven insights, which depend on real-time data collection, analysis, and sharing to understand various aspects of the supply chain. This aids in identifying bottlenecks, predicting demand patterns, optimizing inventory levels, and enhancing operational efficiency (Sharma et al. 2022). Second, automation and digitization involve using robotics, cloud computing, artificial intelligence (AI), and other technologies to automate repetitive processes, boost efficiency, and ensure seamless information flow across the supply chain. Third, collaboration and connectivity are essential, involving stakeholders such as suppliers, manufacturers, distributors, retailers, and customers. This improves collaboration, accelerates response times, and increases visibility across the supply chain. Fourth, flexibility and resilience are key attributes of smart supply chains, designed to be adaptable and responsive to market changes, disruptions, and customer demands. This requires flexible production processes, robust logistics systems, and proactive risk management strategies (Hohenstein 2022).

Integrating open innovation with smart supply chains allows organizations to leverage external knowledge and technological advancements to optimize supply chain operations, develop innovative products and services, and achieve a competitive advantage in the marketplace (Ho et al. 2022; Khan et al. 2022b). By combining collaboration and digital technologies, companies can enhance efficiency, responsiveness, and customer satisfaction. Open innovation supports organizational advancements by providing access to cutting-edge technologies, facilitating collaborative problem-solving, co-creating tailored solutions, making data-driven decisions, and increasing agility in supply chain operations. Embracing open innovation allows companies to utilize digital potential to create smarter, more efficient, and resilient supply chains (Sinha et al. 2020).

In the next part, the study will explore the concepts of OI and SSC and the relationship between the two concepts.

10.3 Open Innovation: Definitions, Dimensions, and Design

Let's go back 320 years to the seventeenth century when six European sea-faring nations embarked on a race to develop accurate maritime navigation and a more precise way to calculate time for determining longitude. This challenge inspired brilliant minds like Galileo Galilei and Christian Huygens to work on solutions. However, it was an English clockmaker in the eighteenth century who ultimately resolved the issue. The innovations from this competition still appear in classic clocks today. This event is an early example of open innovation. Historically, even further back, rulers organized competitions to find methods to improve their realms' welfare. These instances show that while Henry Chesbrough coined the term "open innovation" in 2003 for his book to present a modern view of collaborating with external parties, the underlying concept of open innovation is much older (Isomäki 2018).

Open innovation is a strategy where businesses and other organizations seek and integrate feedback and suggestions from a diverse range of stakeholders, both internal and external (Moradi et al. 2021). This approach involves being transparent about challenges and soliciting input from outside the company. Examples of open innovation include long-term collaborations like that between Delphi and Mobil Eye for developing driverless vehicle technology, co-creation challenges such as the BMW Startup Challenge, and crowdsourcing platforms like "My Starbucks Idea." Unlike the traditional, closed-door approach of conventional commercial R&D, open innovation welcomes contributions from a broader range of stakeholders in problem-solving and product development (Livescault 2023). Essentially, open innovation requires a receptive mindset that values the exchange of information. The concept can be better understood by contrasting it with the more typical "closed innovation" approach.

Chesbrough defines open innovation as "*the use of deliberate transfer of knowledge to speed up internal innovation and broaden markets for external use of innovation.*" In his book, Chesbrough asserts that *open innovation* is a more distributed, participatory, and decentralized approach to innovation as shown in Fig. 10.1. This strategy is a terrific technique for uncovering novel solutions and leveraging

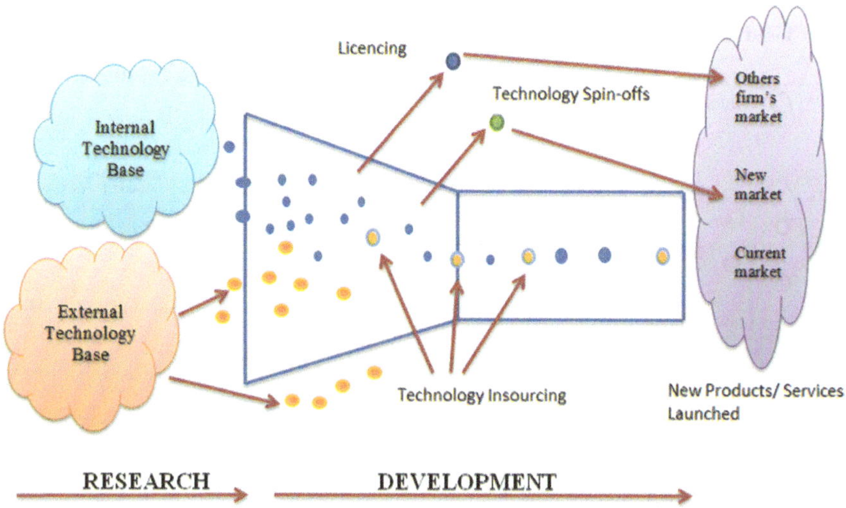

Fig. 10.1 Chesbrough's open innovation. *Source* Author Own

outside information. This decentralized strategy offers several benefits to enterprises (Livescault 2023).

Open versus closed innovation	
Open innovation is based on the premise that brilliant and imaginative individuals are outside the organization	**Closed innovation** assumes that internal skills (ideas) and an iterative management approach can sustainably develop new firms
It achieves its goals and profits from sharing intellectual property. As knowledge grows, decisions are made	The company keeps information to itself and does not share it with any third parties or anyone outside of its four walls
The open innovation funnel on the right looks like a sieve because everyone may participate in the development process. Ideas are multiplying	It is shaped similarly to the funnel on the left, with solid walls representing an internal, controlled, yet safe development process

Both have a place in the world since not all information can be shared with the outside world, but on the other hand, obtaining the feedback of thousands of customers might result in a significant competitive advantage (Attaran and Deb 2018; Isomäki 2018). Figure 10.2 shows the clear description of open and closed innovation.

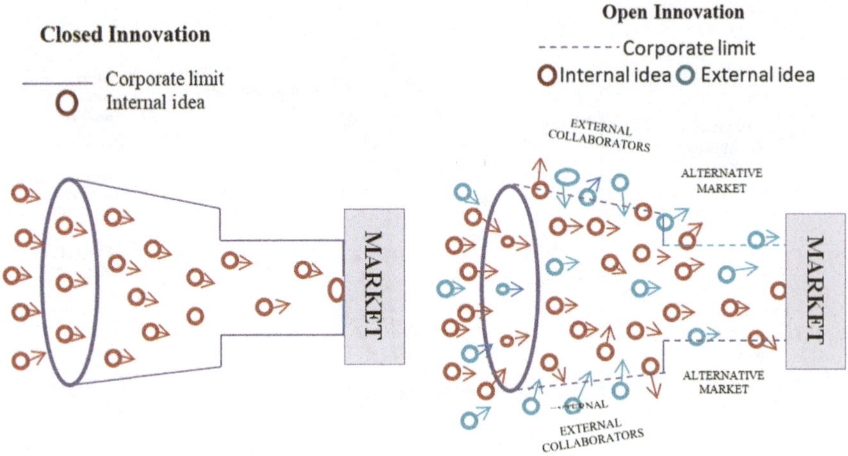

Fig. 10.2 Closed innovation vs. open innovation. *Source* Author Own

10.3.1 Types of Innovation

They are considered to be two types of open Innovation:

- **Inbound Open Innovation**: Inbound innovation aims to discover, select, implement, and assimilate ideas by acquiring knowledge from outside the organization and monitoring the external environment for new data (Cui et al. 2022).
- **Outbound Open Innovation**: Outbound innovation refers to deliberately promoting and capturing internally created innovations in a company's external world. This might be performed by targeting the distribution of a product to critics and journalists or marketing and selling a technology or service to end users to get feedback from those individuals (Briefings 2021).

10.3.2 Sub Types of Open Innovation

Companies may leverage open innovation in various ways, putting the theory into practice whichever best serves their commercial goals. As they consider the various forms it might take, they can acquire a clear grasp of how to do so. Figure 10.3 states the classification of subtype of OI. Open innovation is using four inclusion levels (Alam et al. 2022).

- **Intracompany**—Internal initiatives for open innovation. At the same time, this invention can counter the general tenets of open innovation; intracompany innovation pools resources from several departments.
- **Intercompany**—Open innovation occurs between two (or more) enterprises. A corporate accelerator is the ideal illustration of this kind.

LEVEL OF INCLUSION

PURPOSE OF USE	INTRA-COMPANY	INTER-COMAPNAY	FOR EXPERTS	PUBLICLY OPEN
MARKETING			E.g Conveying information about niche products	E.g Sharing information about a new product
GATHERING INSIGHT	E.g Gathering tacit knowledge from employees	E.g Gathering useful tips from partners	E.g Getting relevant input from the pros in the field	E.g Getting customer feedback to build relevant products
FINDING TALENT	E.g Finding unrecognized talent inside the company	E.g Finding hidden expertise in partnering firms	E.g Recruiting expert from various fields	E.g Finding the best match for the open job position
R & D	E.g Uniting business units for product development	E.g Utilizing collective knowledge between firms	E.g Developing products that require specific expertise	E.g Developing products with the largest possible audience

Fig. 10.3 Subtype of open innovation. *Source* Author Own

- **For experts**: **Innovation is based on open collaboration, which includes the participation of individuals from outside the business who have the necessary** experience and expertise to give valuable advice.
- **Publicly open**—Open innovation welcomes participation from anybody outside the company, regardless of the individual's experience level or reputation.

After determining the level of inclusion, Fig. 10.4 defines the purpose of use as one of the following:

- **Marketing** to spread brand messages and get the word out to the right people.
- **Gathering insight** to learn how to spot and assess relevant data regarding the market and consumer preferences as they currently exist.
- **They were finding talent** to improve the entire skill set of the company's staff and their level of productivity.
- **R&D** to create brand-new goods or services.

There are 16 different forms of open innovation when it compares the extent of inclusion with the use case. Let us imagine, for instance, that your company needed to fill additional roles immediately. You may locate employees with relevant skill sets inside your ranks by using an intracompany strategy for hiring. Use a publicly available methodology to solicit advice from experts and potential consumers throughout development to create goods that appeal to the widest possible audience (Miao et al. 2022). The form of open innovation your company should pick will depend on the degree of inclusion you want to work with and the main objective. You may achieve

	INTRA COMPANY	INTERCOMPANY	FOR PROFESSIONALS	PUBLICLY OPEN
INSIGHTS	Philips	Samsung Philips	Apple U.G.P	Mozilla Nivea
R&D	Facebook Philips	P&G Samsung Philips	Lilly Apple U.G.P	G.A Mozilla G.M Lego Telegram Nivea
MARKETING			Moodle G.A	Lego G.A
TALENT			Facebook	Mozilla G.M G.A

Fig. 10.4 Purpose of use open innovation. *Source* Author Own

significant success by using these criteria as the foundation of your open innovation approach (Richards 2021).

10.3.3 Organization Perception and Open Innovation

Various authors e.g., (Robbins et al. 2021) denote that open innovation describes a new method that has recently existed. It enables organizations to use existing external knowledge rather than beginning their operations from scratch. In this light, developing relationships with other companies, educational institutions, centers of technological innovation, and other information sources is viewed as a positive move. This is what occurs with businesses that utilize open innovation.

(a) The firm works with third parties that are independent knowledge generators.
(b) The Department of Innovation's primary mission is to link and integrate in-house teams with outside knowledge sources.
(c) Internal innovation is still practiced, but its primary focus is on producing information that cannot be obtained anywhere else, provided such knowledge may be sold.

Open Innovation is becoming increasingly popular among organizations of all sizes and regions worldwide to accelerate innovation, reduce risk, and do more with fewer resources. A rocket-like process was employed in models based on OI. Outside input is welcomed throughout both the concept and development phases. These contributions improve the innovation processes and are very beneficial: These outside inputs raise the standard of the ideas produced through open.

Technological development is faster in innovation models, and the risk is smaller due to its connection to the outer world.

Open Innovation is only getting started. There has been a significant movement toward a new paradigm as defined by (Kuhn 2012). Open Innovation is now a common academic topic rather than just the domain of a select few (Del Vecchio et al. 2018). Academics in technology and innovation management developed it, and it is now often mentioned in publications on organizational behavior, general management, and strategy. This contribution is shown in Fig. 10.5.

The question is how far and how long open innovation will last. Real paradigm shifts are permanent and have a longer-lasting effect than hypes in fashion and science (Gassmann et al. 2010).

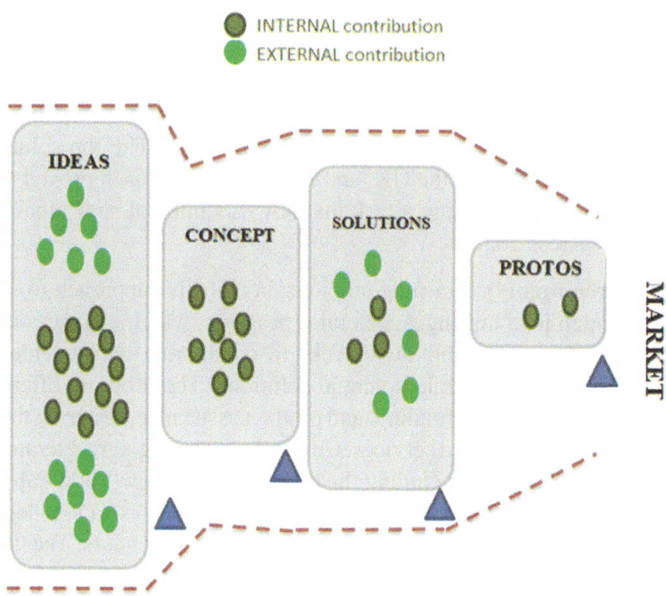

Fig. 10.5 Internal versus external contribution. *Source* Author Own

10.3.4 Examples of Open Innovation

Quick technological advancements have allowed many businesses to experiment in innovative ways over the past ten years. Here are a few instances of open innovation from globally renowned businesses (Richards 2021).

Facebook (Intracompany): With workplace hackathons, the industry leader in social media fosters creativity from the inside out. These events help Facebook bring some of its greatest ideas to the surface from inside its ranks (Ishizaka et al. 2022). By allowing everyone, not just coders, to speak, these events allow staff members to brainstorm and construct early versions of any product ideas they may have before product teams take on them. All Facebook employees have the opportunity to think creatively, and it may even serve as a launching pad for some aspiring businesspeople to pursue new ventures or change careers (Richards 2021).

NASA (Intercompany): Through a process of open innovation, NASA developed a mathematical method to determine the contents of a first aid pack for emergencies, intended for future human-crewed missions. To achieve this, NASA sought to create innovative software and collaborated with Top Coder, Harvard Business School, and the London School of Economics and Political Science. Top Coder users submitted 2833 pieces of code, which helped NASA develop the desired algorithm. Cash awards totaling $24,000 were given to the winners. The successful implementation of open innovation enabled NASA to solve this problem through global collaboration with software enthusiasts and programmers. The top solutions came from Brazil, Indonesia, Japan, the UK, and Japan. This approach proved to be time- and cost-efficient, accomplishing a task that NASA's internal staff alone could not have completed (Alcor 2020).

Samsung (intercompany): Samsung employs an inclusive approach to innovation, particularly through its Samsung Accelerator program, which enhances its external innovation capabilities. This initiative seeks to collaborate with talented individuals and aims to explore multiple potential solutions. The program offers startups' resources such as office space, funding, and product assistance, fostering the development of improved programs and services. Currently, Samsung's Accelerator program operates in several locations, including the tri-city area of New York, Palo Alto, and San Francisco. The Open Innovation Center (OIC) functions like a four-legged stool, naturally extending its collaborations with successful entrepreneurs. The first component is a team of partners, akin to commercial partnerships with other companies. The second component involves venture groups or R&D investments in new businesses, similar to the first approach. The third component is an M&A team that acquires and integrates small teams into Samsung, contributing to product development. The fourth component includes recently established accelerators in Palo Alto and New York City (Alcor 2020).

Coca-Cola (Publicly Open): One of the leading businesses in Innovation is Coca-Cola. The original Coca-Cola fountain machine was designed by Raymond Loewy in 1960. Coca-Cola is using open innovation strategies on every front. The group is on one side, and the other comprises entrepreneurs. The company and its clients are the opponents. Coca-Cola Accelerator is a program that aims to help startup businesses in eight different countries. Berlin, Singapore, Sydney, Buenos Aires, Rio de Janeiro, and Istanbul are just a few of these cities. The San Francisco Bay Area and Bangalore are also covered. These new businesses want to innovate the Happiness Coca-Cola brand in new and exciting ways. The Freestyle machine is an open innovation paradigm case in point. Users from all around the world can combine their preferences. It helps them think of new flavors for Coca-Cola to add to their lineup. The updated product remembers individual tastes. This is done with the help of the official Coca-Cola app. It makes the consumer the focus of manufacturing. (Alcor 2020; Battilana et al. 2022).

UNDP (For Experts): The United Nations Development Program (UNDP) encourages global investment and cooperation for social good. As was already established, the worldwide epidemic has sparked creativity, with specialists from many sectors working nonstop to develop solutions that will benefit the globe in the coming days. The UNDP Accelerator Lab brought medical specialists and the Ministry of ICT and Innovation together to create and deploy anti-epidemic robots as Rwanda struggled to deal with increased cases. At Kigali International Airport, these cutting-edge robots assist with passenger screening, COVID-19 case detection, and extra services like food and medication distribution (Richards 2021).

10.3.5 Open Innovation in Different Expert's Opinions

Open Innovation, in contrast to past studies, "is a paradigm that assumes that firms can and should use external ideas as well as internal ideas… as scholars look to advance their technology". Along with the "new paradigm" idea, Chesbrough also stressed the importance of aligning a firm's commercialization efforts with its business model, whether from internal or external sources (West and Bogers 2014). In the age of open innovation, the benefits of collaboration are growing. The academic community began to emphasize that enterprises should be open to outside innovation as the focus switched from strictly internal R&D efforts (Enkel et al. 2009; Urbinati et al. 2020). "Not all intelligent individuals work for us. Working with clever individuals inside and outside our firm is necessary" (Chesbrough 2006).

10.4 Smart Supply Chain: Definitions and Dimensions

In contrast to the supply chains of yesteryear, which were primarily focused on the availability, transportation, and price of physical assets, today's supply chains are concerned with the management of data, services, and items that are bundled into solutions (Khan et al. 2023). Systems for managing the supply chain today include much more than just deciding where things should be and when they should be done. Supply chain management affects various factors, including the quality of goods and services, delivery timing, expenditures, customer happiness, and profitability. The employment of intelligent technologies in a broad sense allows for data collection from the Internet of Things sensors by smart supply chains. Maintain vigilance and command the supply (Khan et al. 2022a, b, c; Sanders 2022).

The need to integrate business activities across the entire global supply chain, driven by increasing globalization, outsourcing, and information availability, led to the concept of supply chain management in the 1980s and 1990s. This represented a departure from traditional supply chains, which focused only on the most essential logistical and production processes (Ishizaka et al. 2023; Zanoni et al. 2019). As a result of this integration, firms gained improved visibility into the stages preceding and following their own in the chain, prompting each company to focus on optimizing the entire supply chain rather than just their local operations. This shift in focus transformed market competitiveness dynamics. Instead of individual enterprises competing against each other, entire supply chains comprising various businesses began to compete. Consequently, businesses started subcontracting their manufacturing and logistical functions to different companies.

Globalization and technological advancements have enhanced exposure for businesses, improving product traceability and social responsibility programs (Mahmud et al. 2021). The rise of online businesses, the Internet of Things (IoT), and mobile computing have led to new consumer spending habits and operational procedures. The Internet allows customers to interact directly with product distributors, reducing the number of intermediaries in the supply chain and fostering increased collaboration. However, the expectations for speed and delivery times have heightened due to online retailers like Amazon. While standardizing features such as next-day delivery can improve order fulfillment efficiency, it also places additional strain on other parts of the supply chain (Sleight 2021). Employees can now process orders more quickly, necessitating prompt fulfillment. Businesses often order excess materials to avoid shortages, which can lead to waste if customer demand is lower than expected. Technologies such as machine learning, artificial intelligence, and automation have made organizations more responsive to increasing demand (Khan et al. 2021; Shaikh et al. 2022).

In 2017, the typical supply chain had access to fifty times the amount of data it had only five years earlier. However, researchers are only looking at around a quarter of this data. In light of this, it is feasible for the value of essential and timely data, such as information about the weather, unanticipated labor shortages, political turmoil, and transient spikes in demand, to be diminished (Santagate 2018). The COVID-19 pandemic has increased the focus on inventory management and visibility, prompting a shift toward more varied suppliers. After the economic shock from the pandemic has subsided, machine learning and artificial intelligence are expected to play a significant role in diversifying supply change and enhancing the responsiveness and robustness of supply chains. The pandemic may force economies to reorganize their

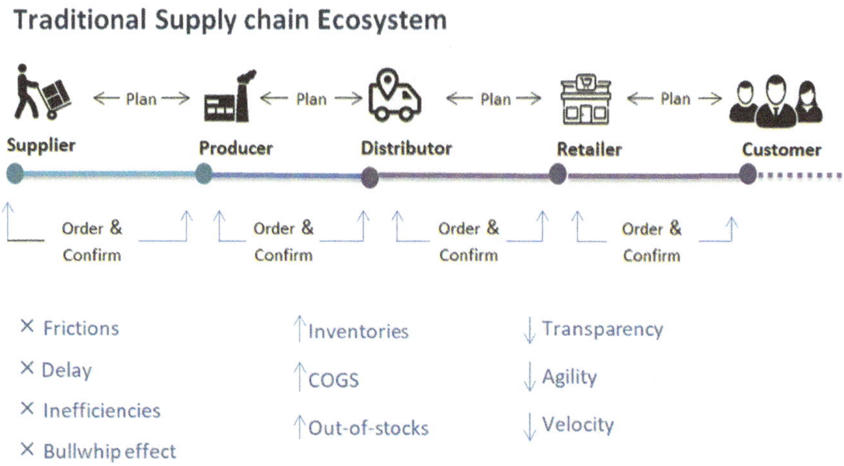

Fig. 10.6 Traditional supply chain ecosystem. *Source* Author Own's

supply chains, moving away from ultra-lean models and prioritizing adaptability and network availability to deliver goods quickly (Lutkevich 2021).

Let us examine the example of a condensed supply chain for bicycles as shown in Fig. 10.6. Before being transformed into bicycle parts, raw materials are initially delivered to original equipment manufacturers (OEMs). These components are then sent to bicycle factories, where staff are assembled to create bicycles (Desjardins et al. 2021). Bicycles are then given to merchants before being sold to customers. These days, only small, local enterprises can find this kind of supply chain simplicity (Foster and Gardner 2022).

Most contemporary supply chain ecosystems have a hundred times greater number of actors. Large-scale supply chain networks must interact with hundreds of OEMs and adhere to various regulatory standards, dramatically increasing network members and greatly complicating operational procedures. Because of this, the contemporary supply chain necessitates a distinct method of interaction and coordination among all partners (Davydov 2022) (Fig. 10.7).

10.4.1 Smart Supply Chain Building Blocks

The development of a smart supply chain relies on integrating essential technologies that drive efficiency, transparency, and optimization across the entire supply chain. The foundation for these advancements begins with a digital backbone, which prepares the environment for more sophisticated tools by enabling digitized workflows. One of the key technologies, the Internet of Things (IoT), plays a crucial role in collecting real-time data about shipment conditions and the state of equipment, improving decision-making. Additionally, artificial intelligence (AI) helps

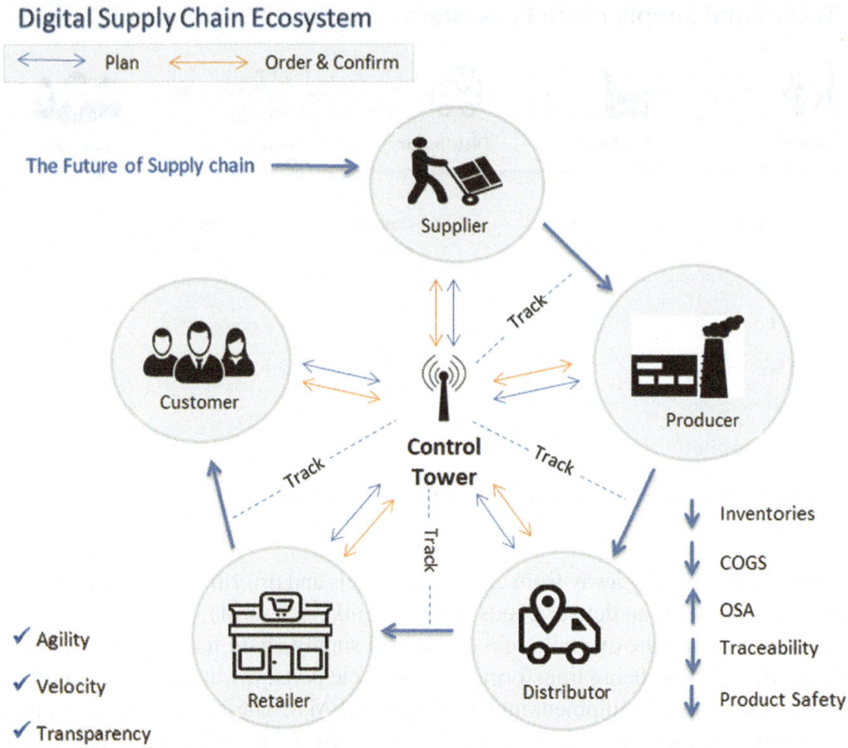

Fig. 10.7 Digital supply chain ecosystem. *Source* Author Own

process large amounts of data to predict demand, streamline planning, and minimize disruptions within the supply chain. To enhance transparency and ensure traceability, blockchain technology is employed, which also facilitates the automated execution of contracts. The supply control tower further boosts visibility, allowing companies to identify issues quickly and optimize supply chain management in real time. Finally, a holistic transformation approach is necessary for companies to harness the full potential of these technologies and achieve a comprehensive smart supply chain transformation (Mubarik and Khan 2024). Table 10.1 outlines these key components, illustrating how each technology contributes to the formation of a smart supply chain.

10.5 Smart Supply Chain and Open Innovation

For the supply chain, open innovation presents only localized difficulties. These days, startups on the market that are quicker, more agile, and more technologically adept are a constant threat to all big companies in the supply chain. Large organizations must always be on the lookout for new open ideas if they want to survive in the face of this

Table 10.1 Key technologies driving smart supply chain transformation

A digital backbone Switching to digitized workflows prepares the ground for the implementation of advanced technologies	**IoT** Allows for the collection of detailed information about shipment conditions and equipment state	**Artificial intelligence** Processes large volumes of data and helps to predict demand, streamline planning, and minimize disruptions
Blockchain Ensures supply chain transparency and traceability and allows for automated execution of contracts	**Supply control tower** It improves visibility across the supply chain, enables quick identification of problems, and optimizes supply chain management	**A holistic transformation approach** It helps the company realize the full potential of a smart supply chain transformation

proliferation of small companies. Employees must collaborate with newcomers in certain situations to create novel solutions or advances. (Dechene 2021). A business model, driven on the concept of open innovation, encourages businesses to seek outside innovation to meet their requirements. Open Innovation speeds up invention by enabling innovators anywhere to take on challenging issues by shattering data silos. This also means that participants in a supply chain will work more closely together in the supply chain sector. First, let us look at what influences the adoption of open supply chain management (Litke et al. 2019).

The industry needs to fundamentally alter its approach to innovation if it wants to break the cycle of low costs and poor margins. Contrary to industrial giants, most startups concentrate on a single element of supply chains. Some of these might affect supply chains' financial results. Businesses collaborate with startups through open innovation programs to improve their supply chain. However, open data exchange with partners along the value chain further reduces ineffective operations (Litke et al. 2019).

Open innovation in smart supply chains offers several pronounced benefits. One significant advantage is the improvement in traceability. A lack of openness and traceability can negatively impact supply networks, leading to waste and the introduction of counterfeit goods by dishonest players. Open innovation addresses these gaps by fostering collaboration among stakeholders. Moreover, advancements such as blockchain technology have further enhanced the effectiveness of these collaborations, ensuring better product quality and minimizing losses (Laukkanen 2023).

Another key benefit of OI is its ability to enable Design for Supply Chain. For instance, the shape of soda cans makes them easy to transport, highlighting the importance of product design in supply chain management. However, the implementation of such designs is often hindered by resistance from various stakeholders. Open Innovation facilitates collaboration between suppliers and manufacturers, allowing for more effective design implementation in the supply chain (Rialti et al. 2022). OI also increases resilience within supply chains. As businesses expand, they often take longer to adapt to environmental changes. Open Innovation makes staff more adaptable, thereby increasing their resistance to change. By sharing information, manufacturers and retailers can become more resilient to disruptions in the supply chain (Sharma et al. 2021). OI contributes to improved sustainability. The sharing of knowledge among stakeholders enables accurate and real-time data reporting, allowing businesses to monitor and reduce emissions throughout their supply chains. Furthermore, innovation encourages shorter supply chains and promotes circularity, both of which enhance sustainability (Litke et al. 2019).

10.6 Open Innovation and Smart Supply Chain: Cases Around the Globe

10.6.1 Procter and Gamble

Procter & Gamble (P&G), one of the world's leading consumer goods companies, is widely recognized for its innovative practices. In the early 2000s, P&G faced increasing pressure to generate new and breakthrough products to maintain its competitive edge in the fast-paced consumer goods industry. To address this challenge, P&G launched its "Connect + Develop" initiative in 2001, embracing open innovation principles to collaborate with external partners. The core idea behind Connect + Develop was to seek innovation from outside the company by connecting with inventors, entrepreneurs, Startups, and researchers globally. Instead of relying solely on internal R&D, P&G sought to leverage the collective creativity and expertise available beyond its organization.

The initiative was a resounding success. P&G was able to access a vast network of external innovators, tapping into a wide range of industries and disciplines. By collaborating with external partners, P&G reduced the time it took to bring new products to market and expanded its product portfolio significantly. One of the most well-known examples of Connect + Develop's success is the creation of Swiffer, a revolutionary cleaning product. P&G partnered with a small company called Techtronic Industries to develop the concept of a new cleaning tool that would be more effective and convenient than traditional methods. This collaboration led to the launch of the Swiffer product line, which became a massive commercial success and transformed the cleaning industry.

Another example of Connect + Develop's impact was the development of Crest Whitestrips, a breakthrough teeth-whitening product. P&G collaborated with a university researcher who had developed a method for teeth whitening using a strip-based application. Through this partnership, P&G was able to commercialize the technology and introduce Crest Whitestrips to the market, revolutionizing at-home teeth-whitening solutions. The Connect + Develop initiative not only accelerated P&G's innovation process but also fostered a culture of openness and collaboration within the company. By embracing open innovation, P&G demonstrated how large corporations can leverage external knowledge and resources to drive transformative and successful product innovations.

This case study highlights how open innovation can enable established companies to access external expertise, leverage diverse perspectives, and introduce innovative products that meet consumers' evolving needs and preferences. It also underscores the importance of cultivating a culture of openness and collaboration within organizations to harness the full potential of external partnerships (Andrişan and Modreanu 2022).

10.6.2 Maersk Line

Maersk Line, the world's largest container shipping company, embarked on a digital transformation journey that integrated smart supply chain principles with open innovation. Facing challenges in the shipping industry, such as inefficiencies, unpredictability, and environmental concerns, Maersk sought to leverage technology and collaboration to optimize its operations.

Smart Supply Chain Implementation: Maersk initiated several smart supply chain initiatives to enhance efficiency and visibility across its operations. They deployed IoT sensors and tracking devices on their containers, vessels, and trucks, enabling real-time cargo monitoring, container conditions, and vessel performance. This data-driven approach improved route optimization, reduced transit times, and minimized potential disruptions, leading to cost savings and better customer service. Furthermore, they implemented AI-powered predictive analytics to optimize container load planning, allowing them to maximize cargo capacity while minimizing empty container movements. Smart warehouse management systems improved inventory accuracy, streamlined operations, and reduced inventory holding costs.

Open Innovation Collaboration: Recognizing the potential of open innovation, Maersk actively sought external collaborations and partnerships. They established the "OceanPro" accelerator program, inviting startups and entrepreneurs to propose innovative solutions for various challenges in the shipping industry. One notable partnership resulted from the OceanPro program with a startup called "LoadME." LoadME developed a digital platform that connected shippers directly with truckers, offering real-time freight bookings and optimizing last-mile deliveries. Maersk integrated this platform into its operations, significantly improving the efficiency of trucking services and enhancing the overall customer experience. The open innovation approach also extended to collaborating with other industry players. Maersk joined the "TradeLens" initiative, a blockchain-powered platform co-developed with IBM, to digitize and secure the end-to-end supply chain process. By sharing data with other shipping companies and stakeholders, Maersk increased transparency and trust, reducing paperwork and simplifying trade processes.

Impact: Integrating smart supply chain technologies and open innovation initiatives significantly benefits Maersk Line. They experience improved operational efficiency, reduced costs, enhanced sustainability by optimizing vessel fuel consumption, and decreased emissions through better route planning. Furthermore, Maersk's customers benefit from improved visibility, more accurate tracking, and streamlined processes. Collaboration with startups and external partners brings fresh perspectives and innovative solutions to long-standing industry challenges.

This case study illustrates how combining smart supply chain technologies and open innovation can transform traditional industries. Maersk Line successfully optimized by leveraging digital tools and partnering with external innovators.

Its operations embraced sustainability and remained competitive in an ever-changing global market (Raza et al. 2023).

The above case study reflects smart supply chains and open innovation in the real world.

The next part will explore the connection between OI and SSC and provide an in-depth discussion of the two concepts.

10.7 Discussion and Implications

Businesses need both internal and external ideas, as well as internal and external channels to market and advance their technology. They can leverage one or both of these resources to their advantage. This is what people refer to when they talk about the "open innovation" approach. Open Innovation merges internal ideas with external contributions to create architectures and systems based on the requirements defined by a business model (Priyono and Hidayat 2022). The rationale for an internally focused, centralized research and development (R&D) strategy is becoming outdated in many sectors today. Immediate application of useful knowledge is necessary. Alongside a company's R&D efforts, integrating external information and ideas forms a new logic of open innovation. This approach embraces external contributions, creating new opportunities to add value (Bereznoy et al. 2021). However, companies still need to undertake the challenging task of transforming promising research discoveries into products and services that meet customer needs. Innovators must collaborate with external parties to pool resources and bring ideas to market swiftly and effectively. This new era of open innovation benefits organizations that can utilize their ideas beyond their current operations and leverage external ideas to enhance their own.

Lucent and Cisco were direct competitors in the same market, but neither business developed similarly. In search of basic insights that may inspire the development of new goods and services, Lucent invested significant resources into investigating the world of cutting-edge components, systems, and materials (Bracci 2022). In contrast, Cisco used a quite different approach in its struggle for the top spot in innovation. Whatever technology the business required, it obtained from outside sources, typically by collaborating with or investing in potential startups (some of which, incidentally, were created by former employees of Lucent). Without doing much of its research, Cisco managed to keep up with the R&D output of perhaps the best industrial R&D organization in the world.

The narrative of Cisco and Lucent is far from unique. IBM's expertise in computing research offered no defense against Intel and Microsoft in the hardware and software industries for personal computers. Similar to how Motorola, Siemens, and other industrial giants stood powerless while Nokia, drawing on its industrial experience from prior decades in the low-tech sectors of rubber boots and wood pulp, shot to the top of wireless telecommunications in under 20 years. Pharmaceutical behemoths like Merck and Pfizer have been observed as some upstarts, such as Genentech, Amgen, and Genzyme, have capitalized on the scientific breakthroughs of others to become significant participants in the biotechnology sector. To be valuable to the organization that developed it, open innovation must be founded on a landscape of freely accessible information. However, a corporation should not limit the information it learns from research to its internal market routes, nor should those internal pathways necessarily be limited to delivering the company's internal expertise to the market (Jovanovic et al. 2019).

Innovation management: One of the newest themes in innovation management is open innovation. This paper aims to examine the conceptual boundaries of open innovation. I answer the what (the open innovation's substance), when (the context dependency), and how (the process) questions in doing so. Open Innovation is a complex idea that may be applied in several ways. One of the least known concepts is the context dependency of open innovation; further study is required to identify the internal and external environment elements that influence performance. The move to open innovation and the numerous open innovation methods are related to the open innovation process. Initial studies are descriptive, based on case studies, and prioritize successful and early adopters as they are with any novel idea. However, not all the lessons the pioneers discovered may apply to the following businesses (Spender et al. 2017).

From open science to open innovation: Open science standards encourage the quick dissemination of the most recent information and enable participation from a wider range of partners in pursuing new knowledge (Jacobsson et al. 2022). This broadens the information, raises its standard, and aids in its dissemination (which starts a new cycle of discovery and diffusion). Although this widespread participation is important, it does not guarantee that scientific information will be effectively commercialized. In some respects, the available scientific principles might make commercializing research more difficult. The idea of open innovation can make it possible for open science's results to be developed and used more quickly. Open innovation presupposes widespread and productive interaction and participation in innovation, much like open science (Plantinga and Adams 2021). However, identifying and creating a business model is also necessary to exploit new information efficiently in open innovation. The business model generates value within the innovation chain, enabling the focus actor to at least partially capture that value. In addition, how intellectual property rights are handled affects how eager and able commercial players are to commit resources and embark on hazardous ventures to create a profitable new process, product, or service. However, too strict IP protection or hasty IP rights assignment at the beginning of a scientific investigation might impede innovation rather than promote it.

10.8 Conclusion

This chapter addresses how organizations might manage the knowledge sources already in supply chains. However, OI stresses the significance of both incoming and outward information flows. Since it incorporates a variety of innovation management strategies, including collaboration in research and development (R&D), market orientation, and supply chain management (SCM) (de Souza et al. 2022), the core notion is not new. For the supply chain, open innovation presents only localized difficulties. These days, startups on the market that are quicker, more agile, and more technologically adept are a constant threat to all big companies in the supply chain. Large organizations must always be on the lookout for new open ideas if they want to survive in the face of this proliferation of tiny companies. In certain situations, they must collaborate with newcomers to create novel solutions or advances (Chatterjee 2020). Intelligent decision-making, self-organization, and responsiveness to environmental fluctuations are a few hallmarks of a "smart" supply chain. Using EDTs, the smart supply chain may become more instrumented, networked, integrated, intelligent, automated, and inventive, enhancing the supply chain's overall performance.

The industry needs to fundamentally alter its approach to innovation if it wants to break the cycle of low costs and poor margins. Contrary to industrial giants, most startups concentrate on a single element of supply chains (Tessaro et al. 2023). Some of these might affect supply chains' financial results. A digital supply chain is a network of connected digital and technological enablers. These enablers simplify supply chain management (SCM) and help organizations better accommodate the fluctuating requirements of their customers. By utilizing various fulfillment techniques, businesses can now provide customers with more choices, speed up the delivery of products, and provide a more individualized level of service. The research finds that automating intelligent supply chain processes may improve efficiency, boost quality, reduce costs and work capital requirements, and boost profits (Das et al. 2021). Conventional supply chains cannot provide these benefits on the same level since they rely on individual silos and human processes. Data-driven intelligent automation techniques, including AI, ML, NLP, and advanced analytics, have simplified planning and performance management in digital supply chains. These technologies also help businesses automate operations, respond rapidly to customer demand and supply shifts, and provide visibility throughout the supply chain. For example, visibility raises responsibility, facilitates the identification of operational glitches, improves vendor efficiency, and boosts the bottom line, ROI, and competitive advantage. Smart supply networks rely heavily on real-time, accurate data (Kayikci et al. 2022).

Supply chain managers may use data and transactional reporting to improve and optimize. A lack of transparency and traceability impacts supply chains (Khan et al 2022b). By encouraging collaboration among stakeholders, open innovation closes these gaps further; recently, developments like blockchain have increased the effectiveness of this cooperation even further. Traceability promotes good product quality while minimizing losses in this manner. Accurate and real-time data reporting is also made possible through stakeholder knowledge sharing. As a result, businesses can track emissions and take action to track and lower emissions throughout their supply chains. However, several factors make it imperative to implement a smart supply chain now more than ever. First, modern supply networks are vulnerable to various threats, from human interference to natural calamities. Improved supply chain readiness, reaction, and recovery capacities are essential for overcoming these disruptions.

In the end, players must forge and deepen relationships, creating more permeable organizational borders and paving the way for competitive advantages that mix internal and external information sources (Egea et al. 2020; Mention 2011). As a result, the notion of open Innovation (OI) was developed since networks and linkages emphasize the interactional nature of inventive performance. Companies benefit from supply chain visibility because it allows them to make well-informed decisions and gives consumers confidence that the items they purchase come from environmentally responsible sources. In recent years, supply chain mapping has emerged as a critical management tool, allowing businesses to absorb shocks better and exploit opportunities.

References

Alam, M. A., Rooney, D., & Taylor, M. (2022). From ego-systems to open innovation ecosystems: A process model of inter-firm openness. *Journal of product innovation management*, *39*(2), 177-201.

Alcor. (2020). *Real World Examples of Open Innovation by Global Organizations*. ALCOR. Retrieved 18 from https://alcorfund.com/insight/real-world-examples-of-open-innovation-by-global-organizations/

Andrişan, G., & Modreanu, A. (2022). Achieving Business Success in the Fourth Industrial Revolution: The Case of Procter & Gamble.

Assis de Souza, T., Alcântara Pinto, G., Rodrigues Antunes, L. G., & Grützmann, A. (2023). SIPOC-OI: a proposal for open innovation in supply chains. *Innovation & Management Review*, *20*(1), 76-93.

Attaran, M., & Deb, P. (2018). Machine learning: the new'big thing'for competitive advantage. *International Journal of Knowledge Engineering and Data Mining*, *5*(4), 277-305.

Battilana, J., Obloj, T., Pache, A.-C., & Sengul, M. (2022). Beyond shareholder value maximization: Accounting for financial/social trade-offs in dual-purpose companies. *Academy of Management Review*, *47*(2), 237-258.

Bereznoy, A., Meissner, D., & Scuotto, V. (2021). The intertwining of knowledge sharing and creation in the digital platform based ecosystem. A conceptual study on the lens of the open innovation approach. *Journal of knowledge Management*, *25*(8), 2022–2042.

Bracci, C. (2022). Management of innovation: Cisco and its A&D strategy.

Briefings, T. O. (2021). *Open Innovation: definition and explanation*. The OR Briefings—Oxcogntia LLC. Retrieved 5 from https://oxford-review.com/oxford-review-encyclopaedia-terms/encyclopaedia-open-innovation-definition-explanation/#

Brunswicker, S., & Vanhaverbeke, W. (2015). Open innovation in small and medium-sized enterprises (SMEs): External knowledge sourcing strategies and internal organizational facilitators. *Journal of small business management*, *53*(4), 1241-1263.

Chatterjee, S. (2020). AI strategy of India: policy framework, adoption challenges and actions for government. *Transforming Government: People, Process and Policy, 14*(5), 757-775.

Chesbrough, H. W. (2006). The era of open innovation. *Managing innovation and change, 127*(3), 34-41.

Chesbrough, H. W., & Vanhaverbeke, W. (2018). Open innovation and Public Policy in the EU with Implications for SMEs. In *Researching open innovation in SMEs* (pp. 455–492). World Scientific.

Chiu, M.-L., & Lin, C.-N. (2022). Developing supply chain open innovation capability: The mediating role of the knowledge creation process, governance mechanism and technology as a driver. *Journal of Innovation & Knowledge, 7*(4), 100264.

Cui, T., Ye, J. H., & Tan, C. H. (2022). Information technology in open innovation: a resource orchestration perspective. *Information & Management, 59*(8), 103699.

Das, D., Datta, A., Kumar, P., Kazancoglu, Y., & Ram, M. (2021). Building supply chain resilience in the era of COVID-19: An AHP-DEMATEL approach. *Operations Management Research*, 1–19.

Davydov, R. (2022). *A guide to making your supply chain smart*. itransition. Retrieved 23 from https://www.itransition.com/blog/smart-supply-chain

de Araújo Burcharth, A. L. L., Lettl, C., & Ulhøi, J. P. (2015). Extending organizational antecedents of absorptive capacity: Organizational characteristics that encourage experimentation. *Technological Forecasting and Social Change, 90*, 269-284.

de Souza, T. A., Pinto, G. A., Antunes, L. G. R., & Grützmann, A. (2022). SIPOC-OI: a proposal for open innovation in supply chains. *Innovation & Management Review, 20*(1), 76-93.

Dechene, H. (2021). *OPEN INNOVATION*. SprintProject. Retrieved 25 from https://www.sprint-project.com/en/expert-opinion/2021/01/open-innovation-definition-challenges-for-supply-chain/

Del Vecchio, P., Di Minin, A., Petruzzelli, A. M., Panniello, U., & Pirri, S. (2018). Big data for open innovation in SMEs and large corporations: Trends, opportunities, and challenges. *Creativity and Innovation Management, 27*(1), 6-22.

Desjardins, E., Apatu, E., Razavi, S. D., Higgins, C. D., Scott, D. M., & Paez, A. (2021). "Going through a little bit of growing pains": A qualitative study of the factors that influence the route choice of regular bicyclists in a developing cycling city. *Transportation research part F: traffic psychology and behaviour, 81*, 431-444.

Egea, M. T. O., Bravo, M. I. R., Moreno, A. R., Domínguez, C. H., & Pullés, D. C. (2020). Technical innovation: trigger or threat for organizational learning? A curvilinear relationship revisited. *Journal of Business & Industrial Marketing, 36*(3), 493-504.

Enkel, E., Bogers, M., & Chesbrough, H. (2020). Exploring open innovation in the digital age: A maturity model and future research directions. In (Vol. 50, pp. 161–168): Wiley Online Library.

Enkel, E., Gassmann, O., & Chesbrough, H. (2009). Open R&D and open innovation: exploring the phenomenon. *R&d Management, 39*(4), 311-316.

Foster, S. T., & Gardner, J. W. (2022). *Managing quality: Integrating the supply chain*. John Wiley & Sons.

Gassmann, O., Enkel, E., & Chesbrough, H. (2010). The future of open innovation. *R&d Management, 40*(3), 213-221.

Ho, W. R., Tsolakis, N., Dawes, T., Dora, M., & Kumar, M. (2022). A digital strategy development framework for supply chains. *IEEE Transactions on Engineering Management*.

Hohenstein, N.-O. (2022). Supply chain risk management in the COVID-19 pandemic: strategies and empirical lessons for improving global logistics service providers' performance. *The International Journal of Logistics Management, 33*(4), 1336-1365.

Ishizaka, A., Khan, S. A., Kheybari, S., & Zaman, S. I. (2022). Supplier selection in closed loop pharma supply chain: A novel BWM–GAIA framework. *Annals of Operations Research*, 1–24.

Ishizaka, A., Khan, S. A., Kheybari, S., & Zaman, S. I. J. A. o. O. R. (2023). Supplier selection in closed loop pharma supply chain: a novel BWM–GAIA framework. *324*(1–2), 13–36.

Isomäki, A. (2018). *Open Innovation—What It Is and How to Do It*. Viima Retrieved 15 from https://www.viima.com/blog/open-innovation

Jacobsson, T. J., Hultqvist, A., García-Fernández, A., Anand, A., Al-Ashouri, A., Hagfeldt, A., . . . Vijayan, A. (2022). An open-access database and analysis tool for perovskite solar cells based on the FAIR data principles. *Nature Energy*, *7*(1), 107–115.

Jalees, T., Qabool, S., Zaman, S. I., & Alam Kazmi, S. H. (2021). Effect of spirituality and ethics on green advertising, and the multi-mediating roles of green buying and green satisfaction. *Cogent Business & Management*, *8*(1), 1920559.

Jovanovic, M., Raja, J. Z., Visnjic, I., & Wiengarten, F. (2019). Paths to service capability development for servitization: Examining an internal service ecosystem. *Journal of Business Research*, *104*, 472-485.

Kayikci, Y., Durak Usar, D., & Aylak, B. L. (2022). Using blockchain technology to drive operational excellence in perishable food supply chains during outbreaks. *The International Journal of Logistics Management*, *33*(3), 836-876.

Khan, M. I., Zaman, S. I., & Khan, S. A. (2023). Relationship and Impact of Block Chain Technology and Supply Chain Management on Inventory Management. In *Blockchain Driven Supply Chain Management* (pp. 53–74). Springer.

Khan, S., Ali, M., & Zaman, I. J. I. J. o. P. M. (2022a). Exploring Interrelationships among Barriers and Enablers of Green Procurement for a Sustainable Supply Chain. *1*(1), 1.

Khan, S. A., Agyemang, M., Ishizaka, A., Zaman, S. I., Ali, S. M., & Laval, J. J. I. j. o. s. e. (2021). Barriers and overcoming strategies to multi-tier sustainable supply chain management: an explorative study in an emerging economy. *14*(6), 1484–1495.

Khan, S. A., Mubarik, M. S., Kusi-Sarpong, S., Gupta, H., Zaman, S. I., & Mubarik, M. (2022b). Blockchain technologies as enablers of supply chain mapping for sustainable supply chains. *Business Strategy and the Environment*, *31*(8), 3742–3756.

Khan, S. A., Mubarik, M. S., Kusi-Sarpong, S., Gupta, H., Zaman, S. I., Mubarik, M. J. B. S., & Environment, t. (2022c). Blockchain technologies as enablers of supply chain mapping for sustainable supply chains. *31*(8), 3742–3756.

Kuhn, T. S. (2012). *The structure of scientific revolutions*. University of Chicago press.

Laukkanen, K. (2023). Identifying and analysing co-operation possibilities between Finland and the Baltics in the circularity of textiles.

Lepore, D., Vecciolini, C., Micozzi, A., & Spigarelli, F. (2023). Developing technological capabilities for Industry 4.0 adoption: An analysis of the role of inbound open innovation in small and medium-sized enterprises. *Creativity and Innovation Management*, *32*(2), 249–265.

Litke, A., Anagnostopoulos, D., & Varvarigou, T. (2019). Blockchains for supply chain management: Architectural elements and challenges towards a global scale deployment. *Logistics*, *3*(1), 5.

Livescault, J. (2023). *Open Innovation*. braineet. Retrieved 19 from https://www.braineet.com/blog/open-innovation#:~:text=Open%20innovation%20is%20the%20practice,business%20for%20solutions%20and%20suggestions.

Lutkevich, B. (2021). *supply chain*. WhatIs.com. Retrieved 21-03-2023 from https://www.techtarget.com/whatis/definition/supply-chain

Mahmud, A., Ding, D., & Hasan, M. M. (2021). Corporate social responsibility: Business responses to Coronavirus (COVID-19) pandemic. *SAGE open*, *11*(1), 2158244020988710.

Masa'deh, R. e., Jaber, M., Sharabati, A.-A. A., Nasereddin, A. Y., & Marei, A. (2024). The Blockchain Effect on Courier Supply Chains Digitalization and Its Contribution to Industry 4.0 within the Circular Economy. *Sustainability*, *16*(16), 7218.

Mention, A.-L. (2011). Co-operation and co-opetition as open innovation practices in the service sector: Which influence on innovation novelty? *Technovation*, *31*(1), 44-53.

Miao, M., Zaman, S. I., Zafar, A., Rodriguez, C. G., & Ali Zaman, S. A. (2022). The augmentation of Knowledge Management through Industry 4.0: case of Aviation sector of emerging economy. *Knowledge Management Research & Practice*, *20*(6), 893–912.

Moradi, E., Jafari, S. M., Doorbash, Z. M., & Mirzaei, A. (2021). Impact of organizational inertia on business model innovation, open innovation and corporate performance. *Asia Pacific Management Review*, *26*(4), 171-179.

Najib, M., Ermawati, W. J., Fahma, F., Endri, E., & Suhartanto, D. (2021). Fintech in the small food business and its relation with open innovation. *Journal of Open Innovation: Technology, Market, and Complexity*, *7*(1), 88.

Naqshbandi, M. M., & Kamel, Y. (2017). Intervening role of realized absorptive capacity in organizational culture–open innovation relationship: Evidence from an emerging market. *Journal of General Management*, *42*(3), 5-20.

Papa, A., Chierici, R., Ballestra, L. V., Meissner, D., & Orhan, M. A. (2021). Harvesting reflective knowledge exchange for inbound open innovation in complex collaborative networks: an empirical verification in Europe. *Journal of knowledge Management*, *25*(4), 669-692.

Plantinga, P., & Adams, R. (2021). Rethinking open government as innovation for inclusive development: Open access, data and ICT in South Africa. *African Journal of Science, Technology, Innovation and Development*, *13*(3), 315-323.

Priyono, A., & Hidayat, A. (2022). Dynamic Capabilities for Open Innovation: A Typology of Pathways toward Aligning Resources, Strategies and Capabilities. *Journal of Open Innovation: Technology, Market, and Complexity*, *8*(4), 206.

Qu, C., & Kim, E. (2024). Reviewing the Roles of AI-Integrated Technologies in Sustainable Supply Chain Management: Research Propositions and a Framework for Future Directions. *Sustainability (2071–1050)*, *16*(14).

Rahmanzadeh, S., Pishvaee, M. S., & Govindan, K. (2022). Emergence of open supply chain management: the role of open innovation in the future smart industry using digital twin network. *Annals of Operations Research*, 1–29.

Raza, Z., Woxenius, J., Vural, C. A., & Lind, M. (2023). Digital transformation of maritime logistics: Exploring trends in the liner shipping segment. *Computers in Industry*, *145*, 103811.

Rialti, R., Marrucci, A., Zollo, L., & Ciappei, C. (2022). Digital technologies, sustainable open innovation and shared value creation: evidence from an Italian agritech business. *British Food Journal*, *124*(6), 1838-1856.

Richards, R. (2021). *Innovation Blog*. MassChallenge. Retrieved 16 from https://masschallenge.org/articles/open-innovation/

Robbins, P., O'Gorman, C., Huff, A., & Moeslein, K. (2021). Multidexterity—A new metaphor for open innovation. *Journal of Open Innovation: Technology, Market, and Complexity*, *7*(1), 99.

Sanders, M. (2022). *Making your supply chain smart*. Binmile. Retrieved 23 from https://binmile.com/blog/a-brief-guide-to-making-your-supply-chain-smart/#Future-of-Supply-Chain-Technology-from-Gartners-perspective

Santagate, S. E. J. (2018). *Supply chain management*. IDC Technology Spotlight. Retrieved 21-3-2023 from https://www.ibm.com/topics/supply-chain-management

Shaikh, A. A., Kumar, A., Jani, K., Mitra, S., García-Tadeo, D. A., & Devarajan, A. (2022). The Role of Machine Learning and Artificial Intelligence for making a Digital Classroom and its sustainable Impact on Education during COVID-19. *Materials Today: Proceedings*, *56*, 3211-3215.

Sharma, M., Luthra, S., Joshi, S., & Kumar, A. (2021). Accelerating retail supply chain performance against pandemic disruption: adopting resilient strategies to mitigate the long-term effects. *Journal of Enterprise Information Management*, *34*(6), 1844-1873.

Sharma, R., Shishodia, A., Gunasekaran, A., Min, H., & Munim, Z. H. (2022). The role of artificial intelligence in supply chain management: mapping the territory. *International Journal of Production Research*, *60*(24), 7527-7550.

Sinha, A., Bernardes, E., Calderon, R., & Wuest, T. (2020). *Digital supply networks: Transform your supply chain and gain competitive advantage with disruptive technology and reimagined processes*. McGraw-Hill Education.

Sleight, T. W. (2021). *Prediction and Hazard Estimation of Polycyclic Aromatic Hydrocarbon Transformation Products* University of Pittsburgh.

Spender, J.-C., Corvello, V., Grimaldi, M., & Rippa, P. (2017). Startups and open innovation: a review of the literature. *European Journal of Innovation Management*, *20*(1), 4-30.

Tessaro, J. A., Harms, R., & Schiele, H. (2023). How startups become attractive to suppliers and achieve preferred customer status: Factors influencing the positioning of young firms. *Industrial Marketing Management, 113*, 100-115.

Urbinati, A., Chiaroni, D., Chiesa, V., & Frattini, F. (2020). The role of digital technologies in open innovation processes: an exploratory multiple case study analysis. *R&d Management, 50*(1), 136-160.

Veldhuizen, M., Blok, V., & Dentoni, D. (2013). Organisational drivers of capabilities for multi-stakeholder dialogue and knowledge integration. *Journal on Chain and Network Science, 13*(2), 107-117.

West, J., & Bogers, M. (2014). Leveraging external sources of innovation: a review of research on open innovation. *Journal of product innovation management, 31*(4), 814-831.

Zanoni, S., Ashourpour, M., Bacchetti, A., Zanardini, M., & Perona, M. (2019). Supply chain implications of additive manufacturing: a holistic synopsis through a collection of case studies. *The International Journal of Advanced Manufacturing Technology, 102*, 3325-3340.

Chapter 11
Smart and Resilient Supply Chains: Role of Smart Supply Chain Mapping

Muhammad Mustafeez ur rehman, Muhammad Shujaat Mubarik, Shuaib Ahmed, and Syed Muzzammil Wasim

11.1 Introduction

In today's volatile business environment, global supply chain (SCs) are more vulnerable than ever to disruptions and shocks that challenge their stability and efficiency (Mubarik and Khan 2024a, b, c). The interconnected nature of supply chains—spanning multiple countries, partners, and stakeholders—exposes them to a wide array of risks, ranging from natural disasters and political instability to economic fluctuations and pandemics. Recent disruption in the form of geo-political conflicts, trade sanctions and natural calamities have exposed vulnerabilities in global supply chains, causing widespread shortages and delays (Kusi-Sarpong et al. 2023). In response to these challenges, digital technologies (DTs) have emerged as potential game changers, offering new ways to improve visibility, flexibility, and responsiveness. Companies that leveraged DTs during the pandemic were able to mitigate the impacts of disruptions by enhancing their capacity to monitor, forecast, and adjust operations in real-time (Mubarik and Khan 2024a, b, c). The use of digital tools not only controlled the immediate crisis but also provided insights into creating more resilient SCs for the future (Jiang et al. 2024).

Despite the increasing interest in digital solutions for supply chain management, the threat of disruptions remains prevalent. There is a growing body of literature on supply chain resilience (e.g., Mubarik et al. 2020; Wieland 2021; Wieland and Durach 2021; Spieske and Birkel 2021; Novak et al. 2021; Shishodia et al. 2023; Ivanov 2024), yet recent crises have revealed that many organizations are still struggling to become truly resilient. This highlights a critical gap: *while technology has provided tools to manage risks more effectively, many organizations have yet to fully embrace*

M. Mustafeez ur rehman · S. Ahmed · S. M. Wasim (✉)
Institute of Business Management, Karachi, Pakistan
e-mail: muzzammil.wasim@iobm.edu.pk

M. S. Mubarik
Edinburgh Business School, Heriot-Watt University, Edinburgh, UK

M. S. Mubarik and S. A. Khan (eds.), *Smart Supply Chain Management*,
https://doi.org/10.1007/978-981-96-1333-5_11

these solutions, and their SCs remain exposed to disruptions. SC resilience—*the ability of a SC to anticipate, prepare for, respond to, and recover from disruptions*—requires organizations to not only respond to immediate shocks but also develop a long-term capability to adapt and thrive in the face of future uncertainties. The three R's (3R) of SC resilience—*readiness, response, and recovery*—are vital components in this regard (Hsieh 2023; Aslam et al. 2020; Queiroz et al. 2022; Yamin 2021).

Among the various strategies available to enhance SC resilience, two capabilities stand out as being underexplored: *supply chain mapping and smart supply chain integration.* Both of these capabilities are essential for organizations looking to build a more resilient supply chain but have not yet been fully harnessed to their potential. Supply chain mapping provides a visual representation of the entire supply chain network, making it easier for firms to identify vulnerabilities, dependencies, and bottlenecks (Chpi et al. 2020; Mibarik et al. 2023). Meanwhile, a smart supply chain leverages advanced technologies such as IoT, AI, machine learning, and blockchain to create a seamless and intelligent integration of all supply chain functions—from procurement and manufacturing to logistics and distribution. By merging these two capabilities, organizations can enhance their real-time visibility and agility, allowing them to respond to disruptions more effectively and efficiently (Azizi et al. 2021).

A smart supply chain (SSC) can be defined as the integration of modern DTs into various supply chain functions, enabling seamless communication, real-time data sharing, predictive analytics, and autonomous decision-making (Pflaum et al. 2021; Yuan et al. 2023; Hofmann et al. 2019).

It allows companies to monitor and manage the flow of goods, services, and information more efficiently and with greater precision (Queiroz et al. 2022). A SSC operates with a high degree of automation, where digital tools and data-driven insights play a central role in optimizing processes, reducing costs, and minimizing risks (Tazhibekova et al. 2023). For example, IoT devices embedded in shipments can track the movement of goods in real-time, while AI-powered analytics can predict potential disruptions before they occur, allowing companies to take proactive measures Wang (2022). This integration creates a more connected, transparent, and agile SC that can adapt to changes quickly and with minimal disruption (Mishra et al. 2024).

Supply chain mapping, on the other hand, is the process of documenting all the entities and processes involved in a SC, from raw material suppliers to end consumers (Choi et al. 2020). Traditionally, supply chain mapping has been a static process, providing a snapshot of the supply chain at a given point in time. However, with the advent of SSCs, supply chain mapping can now become dynamic, thanks to the concept of a digital twin. A digital twin is a real-time virtual model of a physical system, such as a supply chain, which mirrors the state and performance of the system as it evolves (Oliveira and Handfield 2022). By creating a digital twin of a supply chain, organizations can continuously monitor and analyse their supply chain in real-time, enabling them to anticipate and respond to disruptions before they escalate (Sallam 2023).

The ability to create a dynamic supply chain map through a digital twin transforms the way companies manage risks and vulnerabilities (Mubarik 2023; Kusi-Sarpong et al. 2024). Real-time supply chainmapping powered by digital twin technology provides up-to-date information on every component of the supply chain, from suppliers and production facilities to logistics providers and customers. This dynamic mapping capability allows companies to identify potential bottlenecks, alternative routes, and redundant suppliers instantaneously, enabling them to take immediate corrective action. By doing so, organizations can significantly improve their supply chain resilience, ensuring that they can respond to unexpected events with agility and precision (Stephens et al. 2022).

This chapter aims to demonstrate how the integration of smart SCs and digital twin (DT) technology can enhance supply chain resilience by making supply chain mapping dynamic. The primary objective of this chapter is to show how smart supply chains, by providing a digital twin of a firm's supply chain, can make supply chain mapping dynamic, leading to improved supply chain resilience. It also explores how this dynamism can help organizations anticipate risks, adapt to changes in the environment, and recover from disruptions more effectively. The methodology adopted in this chapter is predominantly qualitative, drawing on a review of existing reports and studies that discuss the role of smart supply chain management and supply chain mapping in building resilience. The chapter greatly builds on the work of Mubarik et al. (2023a, b, c), Mubarik and Khan (2024a, b, c), and Choi et al. (2020), among others, who have suggested dynamic supply chain mapping and its impact on supply chain performance. The chapter synthesizes these insights to provide a comprehensive understanding of how organizations can leverage smart SCs and dynamic mapping to achieve resilience.

However, it is important to note that while the integration of digital twin technology with supply chain mapping presents significant opportunities, there are still challenges that need to be addressed. This chapter calls for researchers to undertake feasibility studies to assess the practicalities of implementing digital twin technologies in smart supply chain management. Such studies are essential to determine the technical, financial, and operational challenges that organizations may face in adopting this approach. Without a thorough understanding of the feasibility of these technologies, their potential benefits may remain out of reach.

11.2 Definitions and Dimensions

11.2.1 Smart Supply Chain Management

Over time, advancements in technology have played a crucial role in improving various functions of supply chains, transforming them from simple linear operations into highly interconnected and dynamic systems (Junaid et al. 2023). In the early stages of industrialization, supply chains were primarily manual and localized, with limited data exchange between suppliers, manufacturers, and distributors (Altekar 2023). The advent of the personal computer and basic software systems in the late twentieth century began automating tasks such as inventory management, procurement, and production scheduling (Bhanushali et al. 2024). However, these systems were still fragmented and lacked the ability to provide real-time data and analytics. The introduction of Enterprise Resource Planning (ERP) systems in the 1990s brought about a significant improvement by integrating various business processes, allowing for more cohesive management of supply chain functions (Nyathani et al. 2023). As digital technologies continued to evolve, the introduction of technologies such as the IoT, AI, ML, blockchain, and cloud computing has led to the emergence of what we now refer to as SSC management (Mubarik and Shahbaz 2022).

A SSC is defined as the seamless integration of advanced digital technologies into the various stages of a supply chain to enhance visibility, flexibility, decision-making, and overall efficiency (Shao et al. 2021). According to Ivanov and Dogue (2020), SSC management (SSCM) utilizes technologies such as IoT, AI, and cloud computing to enable real-time data collection and analysis, allowing companies to track and monitor their supply chain processes from end to end. Mubarik et al. (2021a, b) describe SSCM as a technology-driven approach that creates a responsive and agile supply chain capable of adapting to disruptions in real-time. What makes an SSC truly "smart" is its ability to harness the power of data and technology to make informed decisions autonomously, without the need for human intervention at every stage.

In essence, an SSC can be viewed as the sum of a conventional supply chain plus digital technologies. In traditional supply chains, processes are often linear, siloed, and reactive, with limited visibility into the entire chain. Companies in a conventional supply chain might rely on historical data to make predictions about demand or inventory levels, which can result in inefficiencies such as overstocking or stockouts (Wasswa et al. 2023). Moreover, disruptions in traditional supply chains often lead to delays, increased costs, and customer dissatisfaction, as companies are slow to respond to changes in the environment levels (Pane 2024; Reynolds 2024).

By adding digital technologies to the equation, supply chains become more interconnected, data-driven, and agile (Reynolds 2024). IoT devices provide real-time tracking of goods, while AI-powered analytics can predict demand fluctuations and optimize inventory levels Mashat (2024).

Blockchain ensures the transparency and security of data shared across the supply chain, and cloud computing facilitates seamless data sharing among all supply chain partners (Rahayu et al. 2021; Wang and Yang 2022; Xu et al. 2021). This integration of technology transforms a conventional supply chain into an SSC, capable of responding to challenges with speed and accuracy.

SSCs are characterized by several key attributes that differentiate them from traditional supply chains. One of the most important characteristics is real-time visibility. With IoT-enabled devices and sensors embedded throughout the supply chain, SSCs can provide continuous monitoring of goods, inventory levels, and shipments. This level of visibility allows companies to detect potential issues such as delays, stockouts, or quality control problems before they escalate, enabling them to take corrective action in real-time (Njualem 2022). For example, if a shipment is delayed due to weather conditions, an SSC can automatically reroute the delivery, notify the relevant parties, and adjust production schedules, accordingly, minimizing the impact of the disruption.

Another critical feature of SSCs is their ability to leverage predictive analytics. Traditional supply chains typically rely on historical data to make forecasts about demand and supply, but these predictions are often inaccurate and fail to account for sudden changes in the market. SSCs, on the other hand, use AI and machine learning algorithms to analyse vast amounts of real-time data and make more accurate predictions about future trends (Adewusi 2024). By analysing factors such as consumer behaviour, market conditions, and weather patterns, SSCs can anticipate demand fluctuations and adjust inventory levels or production schedules accordingly. This predictive capability helps companies avoid overstocking or stockouts and ensures that products are delivered to customers on time (Oyewole 2024).

Automation is another defining characteristic of SSCs. With advancements in robotics, AI, and machine learning, many processes within the supply chain can now be automated, reducing the need for manual intervention and minimizing the risk of human error. In smart warehouses, for example, robots can be used to pick and pack goods, while automated conveyor systems transport items to the appropriate location. AI-driven systems can also optimize delivery routes, taking into account traffic conditions, fuel costs, and delivery times to ensure that shipments are delivered in the most efficient way possible. This level of automation not only improves operational efficiency but also reduces costs and improves overall supply chain performance (Sodiya 2024).

Collaboration and data sharing are also central to SSCs (Firmansyah and Siagian 2022). In a traditional supply chain, data is often siloed, with each stakeholder working independently and sharing information only when necessary. This lack of transparency can lead to inefficiencies, miscommunication, and delays (Guo 2023). In contrast, SSCs are built on a foundation of collaboration, where data is shared freely and in real-time among all stakeholders, from suppliers and manufacturers to distributors and retailers. Blockchain technology plays a key role in facilitating this transparency, ensuring that data is secure, tamper-proof, and accessible to all relevant parties. By enabling seamless communication and collaboration, SSCs can operate more efficiently, reduce lead times, and improve customer satisfaction (Olutimehin 2024; Wang 2024).

Sustainability is another important feature of SSCs. With increasing pressure on businesses to reduce their environmental footprint, many companies are turning to SSC management to help them achieve their sustainability goals (Aylak 2022; Paul et al. 2022). By using data-driven insights to optimize resource use, reduce waste, and lower energy consumption, SSCs can help companies become more environmentally responsible. For example, AI-powered systems can optimize delivery routes to reduce fuel consumption, while IoT-enabled sensors can monitor energy use in warehouses and production facilities, allowing companies to identify areas where they can reduce energy waste (Olan et al. 2024) (Fig. 11.1).

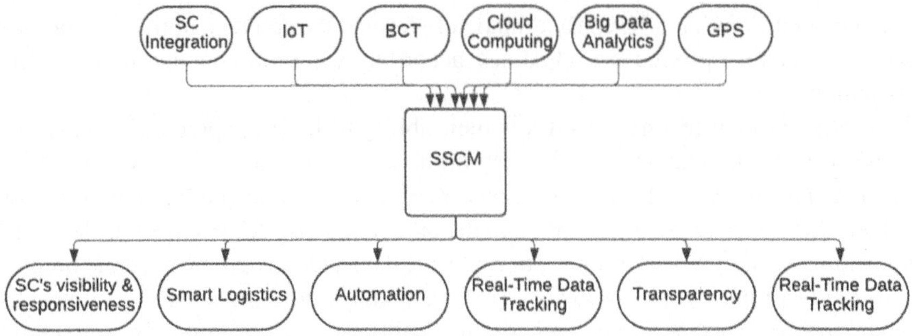

Fig. 11.1 Digital technologies and smart SC

11.2.2 Supply Chain Mapping

SC mapping is a critical tool used to visualize and track the flow of goods, information, and finances across the different stages of a SC, from raw material suppliers to end customers (Ahrisa 2023). It provides a clear representation of how various entities in the SC interact, helping organizations to identify potential risks, inefficiencies, and opportunities for improvement (Purwaningsih 2023). This process has evolved significantly since the 1980s, shaped by technological advancements and growing complexities in global SC s. Initially, SC mapping was static and offered limited visibility into the SC network (Mubarik et al. 2023a, b, c). However, today, with the introduction of dynamic mapping through digital technologies like digital twins, it has become more robust, offering real-time insights and enhancing SC resilience.

In the early stages of SC management, particularly in the 1980s, SC mapping was primarily used to represent linear relationships between suppliers, manufacturers, and distributors (Mubarik et al. 2023a, b, c). The concept was relatively simple, focusing on documenting the movement of goods through the SC. The maps created during this time were often static, providing a snapshot of the SC at a given moment. The limitations of these early maps were apparent—they lacked the ability to show real-time changes in the SC and were not adaptable to disruptions or fluctuations in supply and demand. As a result, while SC mapping provided a useful tool for planning and understanding the flow of materials, it did not offer much in the way of resilience or flexibility (Khan et al. 2022).

As the field of SC management evolved in the 1990s and early 2000s, there was a growing recognition of the importance of SC visibility. Researchers like Stevens (1989) and later, Lambert et al. (1998), contributed to expanding the scope of SC mapping to include upstream, midstream, and downstream processes. Upstream mapping focuses on the network of suppliers and sub-suppliers that provide raw materials and components. Midstream mapping centres on the internal processes of the firm, such as production and inventory management. Downstream mapping involves the coordination of finished goods with customers and distribution networks. This division into three streams helped to provide a more comprehensive view of the

SC, but the maps created were still largely static, providing only a high-level view of the network.

Upstream mapping, in particular, became crucial as organizations sought to gain better visibility into their suppliers' operations (Mubarik et al. 2023a, b, c). This part of the SC involves understanding the flow of raw materials and components from suppliers to manufacturers. In the 1980s and 1990s, upstream mapping was mostly focused on first-tier suppliers, with little attention paid to sub-suppliers or the broader supply network. However, as global SC s became more complex, the need for greater visibility into second tier and even third-tier suppliers became apparent. The lack of visibility in the upstream SC often led to supply disruptions, as firms were unaware of potential risks and bottlenecks in their supplier networks (Wichmann et al. 2020).

Midstream mapping, which focuses on internal processes, also became more sophisticated during this period. Initially, firms used SC maps to document internal processes such as production, warehousing, and distribution (Scholten and Schilder 2015). These maps were useful for identifying inefficiencies and bottlenecks within the organization but did not provide much insight into how the firm interacted with its external partners. Over time, midstream mapping began to incorporate more detailed information about internal operations, including real-time data on production schedules, inventory levels, and transportation logistics. This allowed firms to optimize their internal processes and improve their overall SC performance (MacCarthy et al. 2022).

Downstream mapping involves the coordination of finished goods with customers and distribution networks. In the 1980s, downstream mapping was largely focused on managing relationships with direct customers and distributors. However, as SC s became more customer-centric, firms began to realize the importance of mapping the entire distribution network, including second-tier customers and final consumers. This shift was driven by the need for greater visibility into customer demand and the desire to improve service levels. Downstream mapping now includes real-time data on customer orders, delivery schedules, and transportation routes, enabling firms to respond more quickly to changes in demand and reduce lead times (Mubarik et al. 2023a, b, c).

The static nature of early SC maps was one of their biggest limitations. These maps were often created as part of a strategic planning process and were not updated frequently, making them less useful in managing day-to-day operations. They provided a high-level view of the SC but did not capture the dynamic nature of SC activities. As a result, organizations were often caught off guard by disruptions, such as supplier delays or changes in customer demand. The lack of real-time data and the inability to quickly adapt to changes in the SC limited the effectiveness of static maps in improving SC resilience (Ivanov 2022).

The introduction of DTS have transformed SC mapping from a static process into a dynamic one. Digital twins have revolutionized SC mapping by creating virtual replicas of physical SCs. These digital twins can simulate different scenarios and provide real-time insights into SC performance, allowing firms to anticipate disruptions and respond quickly to changes in the environment. The dynamic nature of modern SC mapping can make it a powerful tool for improving SC resilience. By providing real-time visibility into upstream, midstream, and downstream processes, dynamic SC map can allow organizations to monitor the flow of goods, information, and finances across the entire SC. This can help firms to identify potential risks and bottlenecks before they escalate into major disruptions (Mubarik et al. 2023a, b, c).

11.2.3 Supply Chain Resilience

Resilience, at its core, refers to the ability of a system, organization, or entity to withstand disruptions and quickly recover to a stable state or even adapt and thrive in the face of adversity. This concept has its roots in various disciplines, including ecology, engineering, and psychology, but it is now widely applied in the business and SC contexts. In ecology, resilience is understood as an ecosystem's capacity to absorb shocks and remain functional, or to evolve into a new, stable state after significant changes. In business, resilience extends beyond simply bouncing back from a crisis. It encompasses the ability to foresee disruptions, prepare for them, respond appropriately, and recover in a way that either returns to the previous state or leads to an improved state of functioning (Weiland et al. 2013; Mubarik et al. 2021b).

SC resilience builds on this broader understanding of resilience but applies it to the complex networks of suppliers, manufacturers, distributors, and customers. A resilient SC can handle disruptions—whether they stem from natural disasters, economic fluctuations, or global pandemics—without suffering prolonged operational setbacks. This capability involves both preparing for potential risks and recovering quickly when disruptions occur, ensuring that operations can continue with minimal impact. According to Wieland (2021), SC resilience is not merely about returning to a pre-crisis state; it is about adapting and transforming to thrive in new environments (Wieland and Durach 2021).

The 3R framework—readiness, response, and recovery—captures the key dimensions of SC resilience. Each of these elements plays a crucial role in how well a SC can manage risks and disruptions. The first element, readiness, involves a supply chain's preparedness to face disruptions. This entails the proactive development of strategies, contingency plans, and mechanisms to deal with unforeseen events.

- Readiness is often built through risk assessment, scenario planning, and collaboration with SC partners to ensure that all entities are aligned when a disruption occurs. Companies must evaluate both internal and external risks, identifying weak points in their supply networks and developing plans to mitigate those risks. Intellectual capital, especially structural and relational capital, is essential for building this readiness. Companies with well-established processes, systems, and strong relationships with suppliers and customers are better equipped to handle disruptions, as they can quickly activate their contingency plans and ensure coordination across the supply chain (Mubarik et al. 2021b).

- Response is the second critical component of SC resilience. This refers to the ability of a SC to act swiftly and efficiently when a disruption occurs. Speed is of the essence here, as the longer it takes for a company to respond, the greater the impact of the disruption. An agile SC can adjust production levels, reroute shipments, or switch suppliers rapidly to minimize operational delays. This capability relies heavily on the supply chain's flexibility and adaptability. For example, a company with a diversified supplier base and real-time visibility into its SC operations can quickly pivot when one supplier experiences a delay, finding alternative sources or adjusting inventory levels to meet demand. Moreover, digital technologies such as IoT, AI, and blockchain are increasingly used to enhance response capabilities, enabling real-time monitoring and autonomous decision-making that reduce the time between identifying a disruption and responding to it (Wieland and Durach 2021).
- The final element, recovery, focuses on how quickly and effectively a SC can return to normal operations after a disruption. This phase is critical because prolonged recovery times can lead to significant financial losses, missed opportunities, and damaged relationships with customers and suppliers. Recovery involves more than just restoring operations to their previous state; it also includes learning from the disruption and making necessary improvements to avoid similar issues in the future. Companies that prioritize learning and adaptability in their recovery strategies are more likely to emerge stronger after a crisis, having developed new capabilities and insights that enhance their long-term resilience (Mubarik et al. 2021b; Weiland 2013).

While the 3R framework provides a practical way to understand SC resilience, it is important to recognize that resilience is not a one-size-fits-all concept. The specific strategies a company uses to build readiness, response, and recovery capabilities will depend on its industry, SC structure, and risk profile. For example, in highly regulated industries like pharmaceuticals, companies may need to invest more in compliance and risk mitigation strategies, while industries with less regulation may prioritize agility and speed in their response to disruptions.

The literature on SC resilience emphasizes the importance of both flexibility and robustness. Flexibility refers to the ability of a SC to adapt to changes quickly, while robustness involves designing the SC to withstand shocks without significant changes. A resilient SC must strike a balance between these two elements, ensuring that it can absorb disruptions while also being agile enough to adapt to new conditions. The work of Mubarik et al. (2023a, b, c) further emphasizes the role of intellectual capital in enhancing SC resilience, suggesting that firms with higher levels of human, relational, and structural capital are better positioned to withstand disruptions and recover from them (Mubarik et al. 2021b).

11.2.4 Smart Supply Chain for Supply Chain Mapping: Supply Chain Digital Twin

Digital twin technologies have emerged as a revolutionary tool for enhancing the visibility, agility, and resilience of SC s by offering a virtual representation of the physical SC. This concept is a fundamental pillar of the smart SC, which leverages digital technologies to monitor, predict, and optimize SC activities in real time. A digital twin of a SC provides a dynamic, real-time model of all the players and processes involved in the SC—upstream, midstream, and downstream—offering unprecedented visibility and control over the SC's operations. By continuously synchronizing with its physical counterpart, the digital twin enables firms to simulate scenarios, predict disruptions, and respond proactively, thus transforming SC mapping from a static process to a dynamic, real-time capability (Hribernik et al. 2021).

The digital twin of a smart SC provides a real-time, digital replica of the physical SC, offering firms a comprehensive view of every element involved, from suppliers and manufacturers to distributors and customers. The upstream portion of the SC, which includes the sourcing of raw materials and interactions with suppliers, is often the most complex and opaque. A digital twin in this context enables firms to visualize every supplier and sub-supplier, track the flow of raw materials, and identify potential bottlenecks or risks (Wichmann et al. 2020). For example, if a key supplier is facing a delay due to a shortage of materials, the digital twin can alert the firm in real time, enabling managers to find alternative sources or adjust production schedules accordingly. This level of visibility is critical for mitigating risks and ensuring continuity in SC operations.

In the midstream segment, which includes internal processes such as production, inventory management, and warehousing, a digital twin provides real-time insights into production efficiency, machine performance, and inventory levels. This not only helps firms optimize production schedules but also allows them to anticipate potential disruptions before they occur. For example, if a machine is predicted to fail based on performance data, the digital twin can trigger maintenance activities before the failure occurs, thus avoiding downtime. Additionally, real-time inventory management allows firms to optimize stock levels, reducing the risk of stockouts or overproduction. The ability to simulate various production scenarios also enables firms to test the impact of different decisions, such as changes in production schedules or supplier networks, without disrupting actual operations (Aldrighetti et al. 2021).

Downstream, in the distribution and customer-facing portions of the SC, a digital twin provides visibility into transportation networks, demand patterns, and customer behaviour. Firms can track the movement of goods in real time, optimize delivery routes based on traffic conditions, and ensure that products reach customers on time. This level of real-time visibility into the distribution network helps firms respond quickly to changes in demand, avoid delays, and improve customer satisfaction. By integrating customer demand data into the digital twin, firms can also predict future demand patterns and adjust their SC operations accordingly, ensuring that they can meet customer needs while minimizing waste and excess inventory (Zhang et al. 2022).

One of the most significant advantages of the digital twin is its ability to make SC mapping real-time and dynamic. Traditionally, SC mapping has been a static process, providing a snapshot of the SC at a specific point in time. While this type of mapping can help firms understand the structure of their SC and identify key players, it does not offer the flexibility or visibility needed to manage the dynamic nature of modern SC s. In contrast, a digital twin continuously updates as the physical SC changes, offering a real-time, dynamic view of the SC network. This dynamic mapping capability allows firms to see how changes in one part of the SC impact the rest of the network, enabling them to respond more effectively to disruptions.

For example, if a supplier in the upstream network is affected by a natural disaster, the digital twin can immediately reflect this disruption in the SC map. Firms can then assess the impact of the disruption on production schedules, inventory levels, and customer deliveries, allowing them to make informed decisions about how to mitigate the impact. Similarly, if demand for a product suddenly spikes due to an unexpected event, the digital twin can help firms adjust their production and distribution plans in real time to meet the increased demand. This level of responsiveness is critical for building a resilient and agile SC that can adapt to the rapidly changing market environment.

11.2.4.1 Prerequisites for Implementing Digital Twin in Supply Chain Mapping

To successfully implement digital twin technology and achieve real-time, dynamic SC mapping, several prerequisites must be in place. These prerequisites fall into three main categories: SC mapping capabilities, smart SC technological infrastructure, and organizational readiness. First, firms must have the ability to accurately map their entire SC network, including all suppliers, sub-suppliers, manufacturers, distributors, and customers. This requires a high level of SC transparency and collaboration among all players in the network. Firms must be able to gather and integrate data from multiple sources, including suppliers, logistics providers, and customers, to create a comprehensive view of the supply chain. SC mapping tools that can visualize complex networks and identify relationships between different players are essential for building the foundation of a digital twin. Additionally, firms must establish strong partnerships with their suppliers and other stakeholders to ensure that data is shared in real time, enabling the digital twin to provide an accurate and up-to-date view of the SC (Zhang et al. 2022).

Second, the technological capabilities of the smart SC are critical for supporting the implementation of a digital twin. Several key technologies underpin the functionality of a digital twin. IoT devices, such as sensors and RFID tags, are essential for collecting real-time data on the movement of goods, machine performance, and inventory levels. AI and ML algorithms can analyse this data to identify patterns, predict disruptions, and optimize SC operations. Blockchain technology can provide a secure and transparent way to share data across the SC, ensuring that all stakeholders have access to accurate and tamper-proof information. Cloud computing provides the infrastructure needed to store and process large volumes of data in real time, enabling the digital twin to update continuously as the physical SC changes.

In addition to these technologies, firms must invest in advanced analytics and visualization tools that allow them to interpret the data generated by the digital twin and make informed decisions. These tools should provide real-time insights into SC performance, highlight potential risks or disruptions, and offer recommendations for optimizing operations. For example, a firm might use a dashboard that displays real-time data on inventory levels, production schedules, and delivery times, along with predictive analytics that forecast future demand and identify potential bottlenecks in the SC. By providing a clear and actionable view of the SC, these tools enable firms to respond more effectively to changes in the market environment and improve overall SC performance (West et al. 2021).

Finally, organizational readiness is essential for the successful implementation of a digital twin. This involves not only having the right technological infrastructure in place but also ensuring that the organization has the skills, knowledge, and culture needed to leverage digital twin technology effectively. Firms must invest in training and development programs to ensure that their employees have the skills needed to work with advanced technologies such as AI, IoT, and blockchain. Additionally, firms must foster a culture of collaboration and innovation, where employees are encouraged to experiment with new technologies and approaches to problem-solving.

Organizational readiness also involves aligning the digital twin initiative with the firm's broader business strategy. This requires strong leadership and governance to ensure that the digital twin is integrated into the firm's overall SC management processes and that it is used to support key business objectives, such as improving operational efficiency, reducing costs, and enhancing customer satisfaction. Firms must also establish clear metrics for measuring the success of the digital twin, such as improvements in SC visibility, reductions in lead times, and increases in customer satisfaction. By aligning the digital twin with the firm's strategic goals and ensuring that employees have the skills and knowledge needed to use the technology effectively, firms can maximize the value of their digital twin investment and achieve long-term success.

11.2.5 Smart Supply Chain Mapping for Resilience

Smart SC mapping—*we term it smart because of its dynamic, real-time capabilities powered by digital twin technology*—is continuously updated with real-time data, allowing firms to visualize and respond to changes as they happen. This dynamism is enabled by variety of DTs employed, which together create a comprehensive, real-time replica of the physical SC. This map includes the entire spectrum of the SC—from upstream suppliers and midstream production processes to downstream distribution networks and customer interactions. The *smart* aspect lies in its adaptability, as the map adjusts to new data in real time, providing an ever-evolving view of the SC that helps organizations make informed, proactive decisions (Mubarik and Khan 2024a, b, c).

This transformation of SC mapping is not just a technological leap; it represents a fundamental shift in how organizations manage SC resilience. SC resilience refers to the ability of a SC to anticipate, prepare for, respond to, and recover from disruptions. The three main facets of SC resilience—readiness, response, and recovery—are all improved through smart SC mapping, as it provides organizations with the tools they need to act swiftly and efficiently in the face of challenges.

SSC Mapping and Readiness: The first facet of SC resilience that is enhanced by smart SC mapping is readiness. Readiness refers to an organization's ability to anticipate and prepare for potential disruptions before they occur. With traditional SC mapping, organizations could only plan based on historical data and static representations of their SC networks. This made it difficult to anticipate risks, especially in complex, global SC s where conditions can change rapidly. Smart SC mapping, on the other hand, uses real-time data and predictive analytics to offer a more accurate and up-to-date view of the entire SC. By integrating data from IoT sensors, AI-driven forecasting models, and blockchain-enabled transparency, organizations can identify potential disruptions—such as supplier delays, transportation bottlenecks, or fluctuations in demand—before they escalate into larger problems. For example, smart SC mapping can track inventory levels across different locations in real time, allowing organizations to predict when stock levels are running low and take action to replenish supplies before a shortage occurs. Similarly, predictive analytics can use historical data combined with real-time information to forecast potential disruptions, such as a natural disaster affecting a supplier's operations or a sudden spike in customer demand. This level of foresight enables organizations to develop contingency plans, such as securing alternative suppliers or adjusting production schedules in advance, rather than reacting to disruptions as they occur. This proactive approach to SC management significantly enhances an organization's readiness to face disruptions, ultimately improving overall resilience (Mubarik et al. 2023a, b, c).

SSC Mapping and Response: The second key facet of SC resilience improved by smart SC mapping is response. Response refers to an organization's ability to act quickly and effectively when a disruption does occur. In traditional SC management, the response to disruptions was often slow and reactive because organizations lacked real-time visibility into the SC. Delays in communication between suppliers, manufacturers, and distributors could exacerbate problems, leading to prolonged downtime and increased costs (Kanike 2023). Smart SC mapping solves this problem by providing a real-time, end-to-end view of the SC, allowing organizations to identify the source of a disruption and respond immediately. For example, if a supplier experiences a delay in delivering raw materials, smart SC mapping will reflect this delay in real time, allowing the organization to take immediate action. Managers can quickly assess the impact of the delay on production schedules and customer deliveries, and use the map to explore alternative options, such as rerouting shipments from other suppliers or adjusting production plans. In the past, organizations might not have been aware of the delay until it was too late to take meaningful action, but with smart SC mapping, the disruption is visible as soon as it occurs, and decisions can be made in real time. This speed and agility in responding to disruptions are critical for minimizing the impact on operations and ensuring that the SC continues to function smoothly.

Additionally, smart SC mapping can facilitate better collaboration between SC partners. Because all players in the SC have access to the same real-time data, they can work together more effectively to resolve issues as they arise. For example, if a logistics provider encounters a delay due to bad weather, both the supplier and the manufacturer can see this disruption on the map and coordinate their efforts to mitigate the impact, such as adjusting delivery schedules or rerouting shipments. This level of transparency and collaboration helps organizations respond more efficiently to disruptions, reducing downtime and improving overall SC resilience.

SSC Mapping and Recovery: The third and final facet of SC resilience improved by smart SC mapping is recovery. Recovery refers to an organization's ability to return to normal operations quickly after a disruption and, ideally, to emerge stronger as a result of the experience (Mubarik et al. 2022). Traditional SC mapping provided limited insights into the root causes of disruptions or the most effective strategies for recovering from them. As a result, organizations often took longer to recover from disruptions, and they missed opportunities to learn from these events and improve their future performance. Smart SC mapping changes this by offering a real-time, data-driven approach to recovery. After a disruption, organizations can use the map to analyse the event in detail, identify the root causes, and develop strategies for preventing similar issues in the future. For example, if a factory experiences a production delay due to a machinery breakdown, smart SC mapping can provide insights into the specific machine that failed, the conditions that led to the failure, and the impact of the delay on the rest of the SC. With this information, the organization can take steps to address the root cause of the problem, such as scheduling regular maintenance for the affected machine or investing in more reliable equipment. Moreover, the dynamic nature of smart SC mapping allows organizations to simulate different recovery scenarios and choose the most effective course of action (Al Naimi et al. 2022). For example, managers can use the map to test different strategies for reallocating resources, rerouting shipments, or adjusting production schedules, and see how these changes would affect the overall SC in real time. This capability not only speeds up the recovery process but also helps organizations make more informed decisions about how to adapt and improve their SC operations after a disruption. By learning from each disruption and applying these insights to future operations, organizations can continuously improve their SC resilience over time.

11.2.6 Concluding Remarks

In this chapter, we have explored the transformative potential of digital twin technologies in smart supply chain management, particularly in the context of supply chain resilience. As supply chains become increasingly globalized, complex, and interconnected, organizations face unprecedented challenges in maintaining visibility, efficiency, and agility across their networks. Disruptions, whether in the form of natural disasters, economic fluctuations, or unforeseen pandemics, can severely impact supply chain performance, leading to operational delays, increased costs, and reduced customer satisfaction. Traditional supply chain management approaches, which often rely on static data and linear processes, are no longer sufficient to address these challenges (Naim and Gosling 2023).

Digital twin technology offers a revolutionary solution by providing a real-time, dynamic digital replica of the physical supply chain. Unlike conventional supply chain mapping, which captures a static snapshot of relationships and processes, digital twin technology allows organizations to visualize the entire supply chain—from upstream suppliers and midstream production processes to downstream distributors and customers—in real-time. This dynamic mapping enables firms to monitor, simulate, and optimize supply chain activities continuously, leading to enhanced visibility, responsiveness, and resilience (Mubarik and Khan 2024a, b, c). The key to the effectiveness of digital twin technology lies in its ability to integrate data from various sources, including IoT sensors, ERP systems, and AI-driven analytics. IoT sensors capture real-time data on physical parameters such as location, temperature, and machine performance, while ERP systems provide transactional and historical data on inventory levels, supplier performance, and customer demand. AI algorithms analyze this data to generate insights, identify patterns, and predict potential disruptions (Barja-Martinez et al. 2021). The result is a continuously updated digital twin that mirrors the state of the physical supply chain in real-time, allowing organizations to make data-driven decisions and respond proactively to emerging risks.

Throughout the chapter, we have outlined the process of developing a digital twin, beginning with defining objectives and scope. Organizations must identify key goals, such as improving logistics, optimizing inventory, or enabling predictive maintenance, so the digital twin can address specific needs. The subsequent steps—data collection, model development, testing, and deployment—involve integrating various data sources and simulating real-world behaviours to ensure the model's accuracy. Once deployed, the digital twin provides real-time monitoring and adaptive decision-making, improving operational efficiency and reducing costs. One of the key benefits of digital twin technology is its ability to enhance supply chain resilience across three dimensions: readiness, response, and recovery. Readiness enables firms to anticipate risks and disruptions by using real-time data and predictive analytics to identify vulnerabilities and develop contingency plans in advance. Response improves an organization's ability to act swiftly during disruptions by providing real-time visibility and enabling immediate, data-driven decisions, such as rerouting shipments or adjusting production schedules. Recovery focuses on quickly restoring normal operations after disruptions and learning from the events to prevent future issues. By continuously optimizing processes, digital twins help firms emerge stronger, enhancing long-term resilience (Mubarik and Khan 2024a, b, c).

While the benefits of digital twin technology are clear, its implementation is not without challenges. One of the most significant hurdles in deploying digital twin technology for supply chain mapping is the complexity of modern supply chains. As organizations move beyond their immediate tier-one suppliers, the supply chain becomes increasingly wide, complex, and extended. A single product may involve hundreds, if not thousands, of suppliers, sub-suppliers, and logistics partners spread across multiple geographies. Mapping each node of the supply chain in detail—capturing both the depth (the number of tiers) and the breadth (the number of suppliers at each tier)—is a daunting task that requires significant technological infrastructure, data integration, and collaboration among all supply chain stakeholders (Mubarik et al. 2023a, b, c). Furthermore, the sheer volume of data involved in monitoring a global supply chain in real-time can be overwhelming. IoT sensors generate vast amounts of data every second, and this data must be processed, analyzed, and translated into actionable insights. Integrating data from disparate sources, including suppliers, logistics providers, and customers, can be technically challenging, especially when different stakeholders use different systems and standards for data exchange. Ensuring data accuracy, consistency, and security across the entire supply chain network adds another layer of complexity (Hassija et al. 2020).

Despite these challenges, existing technologies have demonstrated the potential to overcome many of these obstacles. IoT devices, cloud computing, and AI-driven analytics provide the infrastructure needed to collect, process, and analyze vast amounts of data in real-time. Blockchain technology offers a secure and transparent way to share data across the supply chain, ensuring that all stakeholders have access to accurate, tamper-proof information. Advanced analytics and machine learning algorithms can sift through large datasets to identify patterns, detect anomalies, and generate predictive insights. By leveraging these technologies, organizations can build a robust digital twin that accurately reflects the complexity and dynamism of their supply chain.

However, the implementation of digital twin technology for real-time supply chain mapping requires careful planning and coordination. This is where feasibility studies come into play. Feasibility studies are essential for assessing the technical, financial, and operational challenges involved in deploying a digital twin at scale. These studies allow organizations to evaluate the readiness of their supply chain for digital twin technology, identify potential bottlenecks, and develop strategies for addressing these challenges before committing to full-scale implementation. One of the key reasons why feasibility studies are necessary is the need to map the entire supply chain, including upstream suppliers and downstream distributors. As organizations extend their supply chain mapping beyond the first tier of suppliers, the network becomes increasingly complex, with multiple layers of relationships and dependencies. Mapping each node of the supply chain in both depth and breadth requires a significant amount of data collection and integration, as well as collaboration with suppliers, logistics providers, and other stakeholders. Without a clear understanding of the data requirements and technological infrastructure needed to support real-time supply chain mapping, organizations risk investing in digital twin technology without realizing its full potential. Moreover, feasibility studies can help organizations assess the scalability of their digital twin implementation. While digital twin technology may work well for smaller, localized supply chains, scaling it to a global supply chain with hundreds of suppliers and partners requires a different level of infrastructure and capability. Organizations must ensure that their digital twin can handle the volume of data generated by a global supply chain, as well as the complexity of managing multiple tiers of suppliers across different geographies. Feasibility studies allow organizations to test their digital twin at a smaller scale before expanding it to the entire supply chain. However, the implementation of digital twin technology for supply chain mapping is not without challenges. As supply chains become more complex, wide, and extended, mapping each node in depth and breadth becomes increasingly difficult. Feasibility studies are essential for assessing the technical, financial, and operational challenges involved in deploying a digital twin at scale, as well as for ensuring that organizations have the technological infrastructure, data integration, and collaboration needed to support real-time supply chain mapping.

We call upon researchers to further explore this area, particularly the development of scalable solutions for real-time supply chain mapping using digital twin technology. By conducting feasibility studies and investigating the challenges of mapping extended supply chains, researchers can help organizations unlock the full potential of digital twins, enabling them to build more resilient, agile, and efficient supply chains in an increasingly complex and uncertain global market. The future of supply chain management lies in the integration of smart technologies, and digital twins are at the forefront of this transformation. Researchers are encouraged to contribute to this evolving field by addressing the technical and operational challenges associated with real-time supply chain mapping, and by developing innovative solutions that enhance the resilience of global supply chains.

Appendix: Step by Step illustration of building a DT

Digital twins are virtual replicas of physical processes or systems that allow organizations to simulate, analyse, and optimize their operations in real-time. This concept is crucial for improving decision-making, enhancing efficiency, and driving innovation within supply chains. The development of a digital twin follows a structured process that is divided into distinct phases: *Planning, Model Development, and Execution & Optimization.* Each phase contains specific steps that guide the creation and implementation of the digital twin.

Planning Phase

Step 1: Define Objectives and Scope:The first and arguably most critical step in developing a digital twin is to clearly define its objectives and scope. At this stage, the organization must identify the key goals it aims to achieve by implementing a digital twin. These goals could include optimizing inventory levels, improving logistics efficiency, enabling predictive maintenance, or enhancing overall operational transparency. By defining these goals upfront, the organization ensures that the digital twin will focus on specific processes or components within the supply chain. A well-defined scope prevents unnecessary complexity and helps in tailoring the digital twin to address particular pain points or areas for improvement. For example, if the goal is to improve warehouse management, the digital twin would focus on inventory flows and storage processes, integrating data and simulations around these elements to provide valuable insights.

Step 2: Data Collection and Integration: The second step involves gathering and integrating data from various sources across the supply chain. This data forms the foundation of the digital twin and is crucial for ensuring the twin accurately mirrors the real-world supply chain.

There are two primary types of data inputs in this step:

- IoT Sensors: These capture real-time data such as location, temperature, machine status, or movement of products within the supply chain. Sensors placed in transportation vehicles, production equipment, or warehouses continuously monitor physical parameters and feed the digital twin with real-time information.
- ERP Systems: Enterprise Resource Planning (ERP) systems provide transactional and historical data, such as purchase orders, inventory levels, supplier performance, and shipment records. This data enriches the digital twin by giving it context and historical benchmarks, allowing it to simulate future operations based on past and present trends.
- Integrating these two data streams enables the digital twin to offer a holistic view of the supply chain. For instance, it can track the flow of materials from suppliers to warehouses and predict potential delays or bottlenecks based on real-time sensor data and historical trends stored in the ERP system.

Model Development Phase

Step 3: Building the Digital Model: Once the data has been collected and integrated, the next step is to build the digital twin model using simulation and AI techniques. The goal here is to create a virtual representation of the physical processes in the supply chain, ensuring the twin mimics real-world behaviors as closely as possible. The model is designed to simulate various aspects of the supply chain, such as logistics, production processes, inventory management, and demand forecasting. By running simulations, organizations can experiment with different scenarios without impacting real-world operations. For example, the digital twin could simulate the impact of increasing production capacity in one facility or rerouting shipments to another distribution center. AI algorithms also play a role in optimizing these simulations by analyzing vast amounts of data and identifying patterns or trends that human planners might miss.

This step is crucial because the quality of the digital twin depends on how accurately the model replicates real-world processes. If the model is too simplistic, it may not provide actionable insights, whereas a well-calibrated model can offer valuable predictions and optimization opportunities.

Step 4: Testing and Calibration: After the digital model is built, it must be rigorously tested and calibrated to ensure its accuracy. This step is vital for aligning the digital twin with real-world operations. Calibration involves fine-tuning the data inputs, adjusting the model's parameters, and optimizing the algorithms used for machine learning and simulations.

Testing the digital twin involves comparing its outputs with actual data from the physical supply chain. For instance, if the digital twin predicts a delay in shipment, the organization must verify whether that prediction aligns with real-world events. If discrepancies arise, the model may need recalibration, where data inputs or algorithms are adjusted to improve accuracy. The continuous feedback loop between the physical supply chain and the digital twin during this step ensures that the twin remains a reliable and effective tool for decision-making. If errors or mismatches are identified during testing, they are addressed before moving on to the deployment stage.

Execution & Optimization Phase

Step 5: Deployment: Once the digital twin has been tested and calibrated, it is ready for deployment. At this stage, the twin is used for real-time monitoring and feedback. The twin continuously captures data from the physical supply chain and uses this information to update its virtual model in real-time. This real-time monitoring enables organizations to stay ahead of disruptions, respond to operational changes quickly, and continuously improve performance.

For instance, if a machine in the production line begins to show signs of malfunction based on real-time sensor data, the digital twin can predict when the machine is likely to fail and recommend maintenance before a breakdown occurs. Similarly, if customer demand spikes in one region, the digital twin can suggest adjustments to inventory levels or production schedules to meet the increased demand. The deployment phase marks the transition from planning and development to real-world application. The digital twin becomes an integral part of the organization's operational strategy, driving real-time decision-making and enhancing the efficiency and resilience of the supply chain.

Step 6: Continuous Improvement: The final step in the process is continuous improvement. One of the most significant advantages of a digital twin is its ability to evolve and adapt over time. As the physical supply chain changes—whether due to shifts in market demand, new regulations, technological advancements, or emerging risks—the digital twin adjusts accordingly. In this phase, the digital twin supports ongoing improvement efforts by identifying areas where efficiency can be enhanced, costs can be reduced, or risks can be mitigated. For example, by continuously analyzing production data, the twin might identify an opportunity to reduce energy consumption or minimize waste. It can also help the organization prepare for future challenges by simulating potential disruptions and providing insights into how best to respond. This continuous feedback loop allows the organization to remain agile and resilient, making data-driven decisions that enhance overall performance.

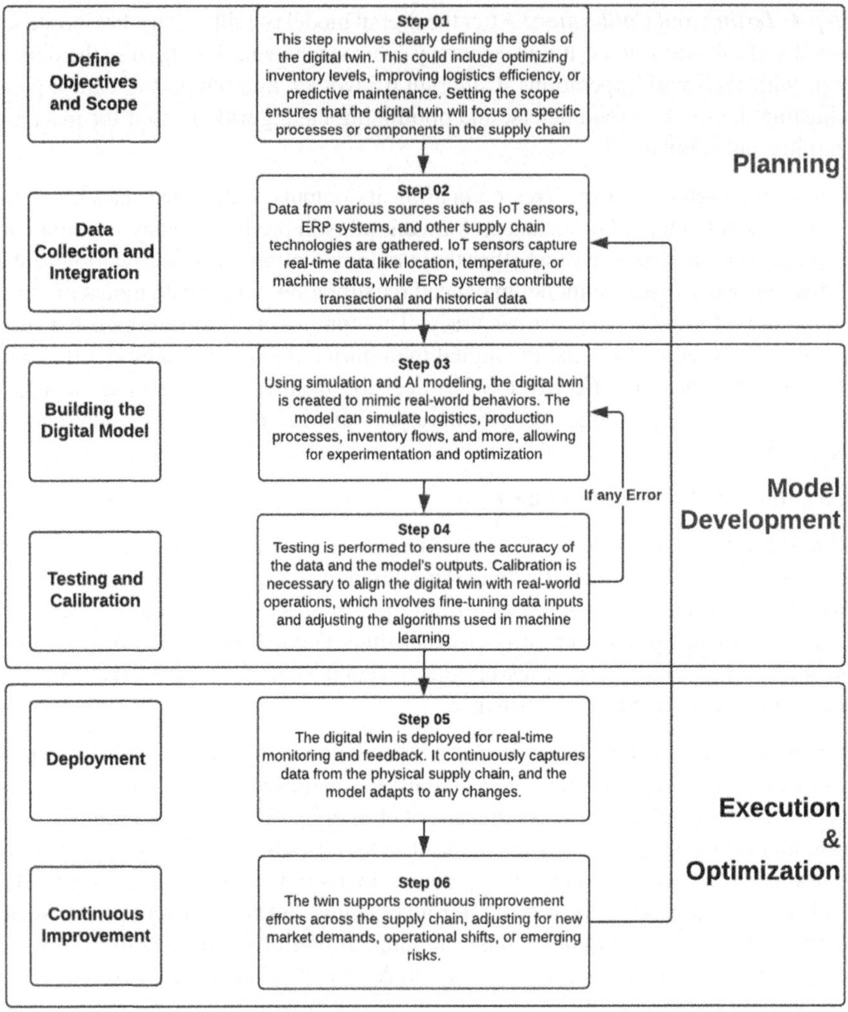

Fig. 11.2 Step by STEP illustration of building a DT

As market conditions change, the twin's insights ensure that the organization can adapt quickly, ensuring long-term success and sustainability (Fig. 11.2).

References

Adewusi, A. (2024). The role of predictive analytics in optimizing supply chain resilience: a review of techniques and case studies. International Journal of Management & Entrepreneurship Research, 6(3), 815–837. https://doi.org/10.51594/ijmer.v6i3.938

Ahrisa, D. (2023). Supply chain flows of sago business in polewali mandar regency, west sulawesi province. E3s Web of Conferences, 454, 02020. https://doi.org/10.1051/e3sconf/202345402020

Al Naimi, M., Faisal, M. N., Sobh, R., & Bin Sabir, L. (2022). A systematic mapping review exploring 10 years of research on supply chain resilience and reconfiguration. International Journal of Logistics Research and Applications, 25(8), 1191-1218.

Aldrighetti, R., Battini, D., Ivanov, D., & Zennaro, I. (2021). Costs of resilience and disruptions in supply chain network design models: a review and future research directions. International Journal of Production Economics, 235, 108103.

Altekar, R. V. (2023). *Supply chain management: Concepts and cases*. PHI Learning Pvt. Ltd.

Aslam, H., Khan, A., Rashid, K., & Rehman, S. (2020). Achieving supply chain resilience: the role of supply chain ambidexterity and supply chain agility. Journal of Manufacturing Technology Management, 31(6), 1185-1204. https://doi.org/10.1108/jmtm-07-2019-0263

Aylak, B. (2022). Impacts of sustainability on supply chain management. European Journal of Science and Technology. https://doi.org/10.31590/ejosat.1075779

Azizi, N., Malekzadeh, H., Akhavan, P., Haass, O., Saremi, S., & Mirjalili, S. (2021). Iot–blockchain: harnessing the power of internet of thing and blockchain for smart supply chain. Sensors, 21(18), 6048. https://doi.org/10.3390/s21186048

Barja-Martinez, S., Aragüés-Peñalba, M., Munné-Collado, Í., Lloret-Gallego, P., Bullich-Massagué, E., & Villafafila-Robles, R. (2021). Artificial intelligence techniques for enabling Big Data services in distribution networks: A review. Renewable and Sustainable Energy Reviews, 150, 111459.

Bhanushali, M. M., Bhardwaj, S., Singh, N. K., Vijayalakshmi, P., Mazumdar, N., & Acharjee, P. B. (2024). From automation to optimization: Exploring the effects of AI on supply chain management. In *Utilization of AI Technology in Supply Chain Management* (pp. 77–94). IGI Global.

Firmansyah, H. and Siagian, H. (2022). Impact of information sharing on supply chain performance through supplier quality management, supply chain agility, and supply chain innovation. Petra International Journal of Business Studies, 5(2), 119-131. https://doi.org/10.9744/ijbs.5.2.119-131

Guo, W. (2023). The impact of blockchain technology on integrated green supply chain management in china: a conceptual study. Journal of Digitainability Realism & Mastery (Dream), 2(02), 58–65. https://doi.org/10.56982/dream.v2i02.112

Hassija, V., Chamola, V., Gupta, V., Jain, S., & Guizani, N. (2020). A survey on supply chain security: Application areas, security threats, and solution architectures. IEEE Internet of Things Journal, 8(8), 6222-6246.

Hofmann, E., Sternberg, H., Chen, H., Pflaum, A., & Prockl, G. (2019). Supply chain management and industry 4.0: conducting research in the digital age. International Journal of Physical Distribution & Logistics Management, 49(10), 945–955. https://doi.org/10.1108/ijpdlm-11-2019-399

Hribernik, K., Cabri, G., Mandreoli, F., & Mentzas, G. (2021). Autonomous, context-aware, adaptive Digital Twins—State of the art and roadmap. Computers in Industry, 133, 103508.

Hsieh, C. (2023). Investigating the role of supply chain environmental risk in shaping the nexus of supply chain agility, resilience, and performance. Sustainability, 15(20), 15003. https://doi.org/10.3390/su152015003

Ivanov, D. (2020) Predicting the Impacts of Epidemic Outbreaks on Global Supply Chains: A Simulation-Based Analysis on the Coronavirus Outbreak (COVID-19/ SARS-CoV-2) Case. Transportation Research Part E: Logistics and Transportation Review, 136, Article ID: 101922. https://doi.org/10.1016/j.tre.2020.101922

Ivanov, D. (2022). Lean resilience: AURA (Active Usage of Resilience Assets) framework for post-COVID-19 supply chain management. The International Journal of Logistics Management, 33(4), 1196-1217.

Ivanov, D. (2024). Two views of supply chain resilience. *International Journal of Production Research*, *62*(11), 4031-4045.

Jiang, H., He, B., Mubarik, M. S., & Shi, S. (2024). Role of supply chain digitalization and global supply chain in decarbonization of natural resources sector supply chain. *Journal of environmental management, 370*, 122689. Advance online publication. https://doi.org/10.1016/j.jenvman.2024.122689

Junaid, M., Zhang, Q., Cao, M., & Luqman, A. (2023). Nexus between technology enabled supply chain dynamic capabilities, integration, resilience, and sustainable performance: An empirical examination of healthcare organizations. *Technological forecasting and social change, 196*, 122828.

Kanike, U. K. (2023). Factors disrupting supply chain management in manufacturing industries. Journal of Supply Chain Management Science, 4(1-2), 1-24.

Khan, S., Mubarik, M., Kusi-Sarpong, S., Gupta, H., Zaman, S., & Mubarik, M. (2022). Blockchain technologies as enablers of supply chain mapping for sustainable supply chains. Business Strategy and the Environment, 31(8), 3742-3756. https://doi.org/10.1002/bse.3029

Kusi-Sarpong, S., Mubarik, M. S., & Khan, S. A. (2024). Supply chain mapping for "visilience": Role of blockchain-driven supply chain management. In *The Palgrave Handbook of Supply Chain Management* (pp. 585-599). Cham: Springer International Publishing.

Lambert, D.M., Cooper, M.C. and Pagh, J.D. (1998), "Supply Chain Management: Implementation Issues and Research Opportunities", The International Journal of Logistics Management, Vol. 9 No. 2, pp. 1-20. https://doi.org/10.1108/09574099810805807

MacCarthy, B. L., Ahmed, W. A., & Demirel, G. (2022). Mapping the supply chain: Why, what and how?. International Journal of Production Economics, 250, 108688.

Mashat, R. (2024). Impact of internet of things adoption on organizational performance: a mediating analysis of supply chain integration, performance, and competitive advantage. Sustainability, 16(6), 2250. https://doi.org/10.3390/su16062250

Mishra, N. K., Pande Sharma, P., & Chaudhary, S. K. (2024). Redefining agile supply chain practices in the disruptive era: a case study identifying vital dimensions and factors. Journal of Global Operations and Strategic Sourcing.

Mubarik, M. (2022). Effect of Blockchain Based Supply Chain Management on Supply Chain Resilience: Mediating Role of Supply Chain Mapping and Channel Integration (Doctoral dissertation, Universiti Tun Hussein Onn (Malaysia)).

Mubarik, M. S. (2023). A firm-based perspective of the notion of "carbon neutrality": The role of supply chain mapping. In *Recent developments in green finance, green growth and carbon neutrality* (pp. 63–84). Elsevier.

Mubarik, M. S., & Khan, S. A. (2024a). An Economic Perspective of DSCM Adoption. In The Theory, Methods and Application of Managing Digital Supply Chains (pp. 59–76). Emerald Publishing Limited.

Mubarik, M. S., & Khan, S. A. (2024b). Disruptive Digital Technologies and Contemporary Supply Chains. In *The Theory, Methods and Application of Managing Digital Supply Chains* (pp. 15–39). Emerald Publishing Limited.

Mubarik, M. S., & Khan, S. A. (2024c). Future of Digital Supply Chain Management. In The Theory, Methods and Application of Managing Digital Supply Chains (pp. 163–178). Emerald Publishing Limited.

Mubarik, M. S., Bontis, N., Mubarik, M., & Mahmood, T. (2022). Intellectual capital and supply chain resilience. *Journal of intellectual capital, 23*(3), 713-738.

Mubarik, M. S., Khan, S. A., Kusi-Sarpong, S., Brown, S., & Zaman, S. I. (2023a). Supply chain mapping, sustainability, and industry 4.0. Taylor & Francis.

Mubarik, M. S., Kusi-Sarpong, S., Govindan, K., Khan, S. A., & Oyedijo, A. (2023). Supply chain mapping: a proposed construct. International Journal of Production Research, 61(8), 2653-2669.

Mubarik, M. S., Naghavi, N., Mubarik, M., Kusi-Sarpong, S., Khan, S. A., Zaman, S. I., & Kazmi, S. H. A. (2021). Resilience and cleaner production in industry 4.0: Role of supply chain mapping and visibility. Journal of cleaner production, 292, 126058.

Mubarik, M., Khan, S., Acquaye, A., & Mubarik, M. (2023c). Supply chain mapping for improving "visilience": a hybrid multi-criteria decision making based methodology. Journal of Multi-Criteria Decision Analysis, 30(5-6), 173–189. https://doi.org/10.1002/mcda.1807

Mubarik, Muhammad & Bontis, Nick & Mubarik, Mobashar & Mahmood, Tarique. (2021b). Intellectual capital and supply chain resilience. Journal of Intellectual Capital.

Mubarik, Muhammad & Devadason, Evelyn & Govindaraju, V.G.R Chandran. (2020). Human capital and export performance of small and medium enterprises in Pakistan. International Journal of Social Economics. ahead-of-print. https://doi.org/10.1108/IJSE-03-2019-0198.

Naim, M. M., & Gosling, J. (2023). Revisiting the whole systems approach: designing supply chains in a turbulent world. The International Journal of Logistics Management, 34(1), 5-33.

Njualem, L. (2022). Leveraging blockchain technology in supply chain sustainability: a provenance perspective. Sustainability, 14(17), 10533. https://doi.org/10.3390/su141710533

Novak, D. C., Wu, Z., & Dooley, K. J. (2021). Whose resilience matters? Addressing issues of scale in supply chain resilience. *Journal of Business Logistics*, *42*(3), 323-335.

Nyathani, R., Allam, K., & Engineer, B. I. (2023). Synergizing AI, Cloud Computing, and Big Data for Enhanced Enterprise Resource Planning (ERP) Systems. *Joseph, Sb, Samon Daniel, and GO Godwin. " The Evolving Relationship Between Computers and Society, Impacts, Challenges, and Opportunities.*

Olan, F., Arakpogun, E., Jayawickrama, U., Suklan, J., & Liu, S. (2024). Sustainable supply chain finance and supply networks: the role of artificial intelligence. IEEE Transactions on Engineering Management, 71, 13296-13311. https://doi.org/10.1109/tem.2021.3133104

Oliveira, M. and Handfield, R. (2022). The role of an open analytics culture and analytics skills in adopting real-time supply chain analytics for higher performance. The International Journal of Logistics Management, 34(6), 1833-1857. https://doi.org/10.1108/ijlm-03-2022-0108

Olutimehin, D. (2024). Optimizing fmcg supply chain dynamics: a novel framework for integrating operational efficiency and customer satisfaction. International Journal of Management & Entrepreneurship Research, 6(3), 770–779. https://doi.org/10.51594/ijmer.v6i3.934

Oyewole, A. (2024). Reviewing predictive analytics in supply chain management: applications and benefits. World Journal of Advanced Research and Reviews, 21(3), 568–574. https://doi.org/10.30574/wjarr.2024.21.3.0673

Pane, M. (2024). The effect of supply chain disruptions on business post covid-19. https://doi.org/10.31219/osf.io/6zhq2

Paul, S., Ali, S., Hasan, M., Paul, S., & Kabir, G. (2022). Critical success factors for supply chain sustainability in the wood industry: an integrated pca-ism model. Sustainability, 14(3), 1863. https://doi.org/10.3390/su14031863

Pflaum, A., Bodendorf, F., Prockl, G., & Chen, H. (2021). Introduction to the minitrack on the digital supply chain of the future: applications, implications, business models. https://doi.org/10.24251/hicss.2021.577

Purwaningsih, T. (2023). Training on identification of business units of village-owned enterprises (bumdes). iccs, 1(2). https://doi.org/10.18196/iccs.v1i2.157

Queiroz, M., Wamba, S., Jabbour, C., & Machado, M. (2022). Supply chain resilience in the uk during the coronavirus pandemic: a resource orchestration perspective. International Journal of Production Economics, 245, 108405. https://doi.org/10.1016/j.ijpe.2021.108405

Rahayu, S., Sharmelen, A., Vasanthan, L., Azahari, A., & Chai, J. (2021). Military supply chain management and blockchain development. https://doi.org/10.5121/csit.2021.111608

Reynolds, S. (2024). Examining the impact of digital transformation on supply chain processes. https://doi.org/10.20944/preprints202406.0221.v1

Sallam, K. (2023). Internet of things (iot) in supply chain management: challenges, opportunities, and best practices. Sustain. Mach. Intell. J., 2. https://doi.org/10.61185/smij.2023.22103

Scholten, K. and Schilder, S. (2015). The role of collaboration in supply chain resilience. Supply Chain Management an International Journal, 20(4), 471-484. https://doi.org/10.1108/scm-11-2014-0386

Shao, X. F., Liu, W., Li, Y., Chaudhry, H. R., & Yue, X. G. (2021). Multistage implementation framework for smart supply chain management under industry 4.0. *Technological Forecasting and Social Change, 162*, 120354.

Shishodia, A., Sharma, R., Rajesh, R., & Munim, Z. H. (2023). Supply chain resilience: A review, conceptual framework and future research. *The International Journal of Logistics Management, 34*(4), 879-908.

Sodiya, E. (2024). Ai-driven warehouse automation: a comprehensive review of systems. GSC Advanced Research and Reviews, 18(2), 272–282. https://doi.org/10.30574/gscarr.2024.18.2.0063

Spieske, A., & Birkel, H. (2021). Improving supply chain resilience through industry 4.0: A systematic literature review under the impressions of the COVID-19 pandemic. *Computers & Industrial Engineering, 158*, 107452.

Stephens, A., Kang, M., & Robb, C. (2022). Linking supply chain disruption orientation to supply chain resilience and market performance with the stimulus–organism–response model. Journal of Risk and Financial Management, 15(5), 227. https://doi.org/10.3390/jrfm15050227

Stevens, G.C. (1989), "Integrating the Supply Chain", International Journal of Physical Distribution HYPERLINK "https://www.emerald.com/insight/publication/issn/0269-8218"& HYPERLINK "https://www.emerald.com/insight/publication/issn/0269-8218" Materials Management, Vol. 19 No. 8, pp. 3–8. https://doi.org/10.1108/EUM0000000000329

Tazhibekova, K., Shametova, A., Maharramov, R., & Makar, S. (2023). The role of information and digital tools in supply chain management during the covid crisis. Acta Logistica, 10(1), 25–34. https://doi.org/10.22306/al.v10i1.344

Wang, H. (2022). Linking ai supply chain strength to sustainable development and innovation: a country-level analysis. Expert Systems, 41(5). https://doi.org/10.1111/exsy.12973

Wang, M. and Yang, Y. (2022). An empirical analysis of The supply chain flexibility using blockchain technology. Frontiers in Psychology, 13. https://doi.org/10.3389/fpsyg.2022.1004007

Wang, Y. (2024). A quantitative study of the impact of supply chain management on customer satisfaction. Frontiers in Business Economics and Management, 12(3), 285–290. https://doi.org/10.54097/vteg6mls

Wasswa, J., Oundo, H., Oteba, M., Komakech, H., Ochola, I., Mwebaze, S., … & Lugada, E. (2023). Leveraging electronic logistics management information systems to enhance and optimize supply chain response during public health emergencies: lessons from covid-19 response in uganda. Journal of Pharmaceutical Policy and Practice, 16(1). https://doi.org/10.1186/s40545-023-00517-4

Weiland, J. E., Littke, W. R., & Haase, D. L. (2013). Forest nurseries face critical choices with the loss of methyl bromide fumigation. California Agriculture, 67(3), 153-161. https://doi.org/10.3733/ca.v067n03p153

Wieland, A. (2021). Dancing the supply chain: Toward transformative supply chain management. *Journal of Supply Chain Management, 57*(1), 58-73.

Wieland, A., & Durach, C. F. (2021). Two perspectives on supply chain resilience. *Journal of Business Logistics, 42*(3), 315-322.

Xu, P., Lee, J., Barth, J., & Richey, R. (2021). Blockchain as supply chain technology: considering transparency and security. International Journal of Physical Distribution & Logistics Management, 51(3), 305-324. https://doi.org/10.1108/ijpdlm-08-2019-0234

Yamin, M. (2021). Investigating the drivers of supply chain resilience in the wake of the covid-19 pandemic: empirical evidence from an emerging economy. Sustainability, 13(21), 11939. https://doi.org/10.3390/su132111939

Yuan, Y., Tan, H., & Liu, L. (2023). The effects of digital transformation on supply chain resilience: a moderated and mediated model. Journal of Enterprise Information Management, 37(2), 488-510. https://doi.org/10.1108/jeim-09-2022-0333

Zhang, G., MacCarthy, B. L., & Ivanov, D. (2022). The cloud, platforms, and digital twins—Enablers of the digital supply chain. In The digital supply chain (pp. 77–91). Elsevier.

Chapter 12
Smart Warehousing in FMCG Sector: Challenges and Remedies

Sherbaz Khan, Syed Imran Zaman, Altaf Ahmed Shiekh, and Munazza Saeed

12.1 Introduction

The advent of smart warehousing has revolutionized the global supply chain landscape, making it an integral component of smart supply chain management. By leveraging cutting-edge technologies, smart warehousing has redefined the traditional roles of warehouses from being cost centers to becoming value centers (Zhen and Li 2022; Ali et al. 2024; Mubarik and Khan 2024). This transformation has enabled organizations to optimize their storage and distribution processes, minimize operational inefficiencies, and enhance overall supply chain resilience. With smart warehousing, companies can achieve real-time visibility, automate routine tasks, and integrate operations seamlessly with upstream and downstream supply chain activities, thereby creating a more dynamic and agile ecosystem. Globally, the impact of smart warehousing has been nothing short of disruptive. For instance, Amazon's fulfillment centers have become a benchmark for smart warehousing, utilizing robotics and machine learning to process orders with unparalleled efficiency and accuracy. Similarly, companies like Alibaba and JD.com in China have implemented fully automated warehouses that can operate 24/7, drastically reducing human errors and improving speed-to-market. In Europe, DHL's Smart Warehouse Solutions utilize IoT-enabled sensors and predictive analytics to optimize inventory levels and reduce waste. These examples underscore how smart warehousing has become a game-changer, driving efficiency and sustainability in supply chains worldwide (Allison 2023).

S. Khan (✉) · S. I. Zaman
Department of Business Administration, Jinnah University for Women, Karachi, Pakistan
e-mail: analyzeus@gmail.com

A. A. Shiekh
Government of Pakistan, Islamabad, Pakistan

M. Saeed
FAST-National University, Lahore Campus, Islamabad, Pakistan

M. S. Mubarik and S. A. Khan (eds.), *Smart Supply Chain Management*,
https://doi.org/10.1007/978-981-96-1333-5_12

Despite these advancements, certain sectors face unique challenges that make warehousing a critical area of focus. Among them, the fast-moving consumer goods (FMCG) sector stands out due to its vast product variety and high turnover rates. FMCG companies often grapple with issues related to warehousing operations, given the sheer diversity and volume of their products. The complexities of managing perishable goods, diverse packaging requirements, and the need for rapid order fulfillment often strain traditional warehousing systems. Furthermore, the FMCG sector's reliance on just-in-time inventory systems adds pressure to maintain accuracy and timeliness in warehouse operations. Key issues in FMCG warehousing include inefficiencies in picking and binning processes, which can lead to order inaccuracies and delays. Inventory management is another significant challenge, particularly in handling expiry-sensitive products, where delays or errors can result in wastage and financial losses (Tiwari 2023; Raman et al. 2023; Kukard 2023). Pilferage and shrinkage are additional concerns, especially in regions with weaker regulatory frameworks. The demand for rapid replenishment, coupled with fluctuating consumer preferences, makes it imperative for FMCG companies to adopt warehousing solutions that are both robust and flexible (Mahdzir et al. 2023).

In response to these challenges, FMCG companies worldwide are increasingly exploring smart warehousing as a solution. By adopting technologies such as automated guided vehicles (AGVs), smart shelves, and AI-powered analytics, businesses aim to streamline their operations and address persistent issues in inventory management, order fulfillment, and resource optimization (Li and Huang 2024). For example, Nestlé's implementation of a smart warehouse in the UAE integrates robotic picking systems with real-time data analytics, reducing operational costs while improving efficiency. Similarly, Unilever's use of AI-driven warehousing systems in India has enhanced inventory accuracy and minimized waste.

However, the adoption of smart warehousing has not always yielded the desired results. In some cases, the transition has been counterproductive, leading to operational bottlenecks and increased costs. For instance, a European FMCG company faced challenges in integrating its legacy systems with smart warehousing technologies, resulting in data inconsistencies and workflow disruptions. Similarly, a multinational food and beverage company in Asia experienced delays in implementation due to insufficient training and resistance from employees, highlighting the human factors that can hinder the effectiveness of smart warehousing. These examples illustrate that the barriers to smart warehousing adoption extend beyond technological challenges. Issues such as high initial investment costs, resistance to change, lack of skilled workforce, and regulatory constraints often impede successful implementation. This chapter aims to address these barriers by exploring the adoption challenges specific to the FMCG sector. While there is substantial literature on the performance of smart warehousing, this chapter narrows its focus to the barriers affecting adoption, thereby filling a critical gap in existing research.

The context of this study is the FMCG sector in Pakistan, a region characterized by its unique challenges and opportunities. The FMCG industry in Pakistan is one of the fastest-growing sectors, contributing significantly to the country's economy

(Dantas et al. 2023). However, it faces distinct logistical challenges, including fragmented supply chains, inadequate infrastructure, and regulatory hurdles. The sector's reliance on manual and semi-automated warehousing systems further exacerbates inefficiencies, making it an ideal case for studying smart warehousing adoption. Moreover, Pakistan's rapidly evolving consumer market, coupled with increasing competition, necessitates the adoption of innovative solutions to stay competitive. To examine these issues, data was collected from 49 experts in the FMCG sector using a three-point Likert scale *(not important, somewhat important, and important).* These experts, including supply chain managers, warehouse operators, and logistics consultants, provided insights into the perceived importance of various barriers to smart warehousing adoption. The data highlights the critical challenges faced by the industry, offering a nuanced understanding of the factors that influence adoption decisions.

This study contributes to the growing body of knowledge on smart warehousing by providing a comprehensive analysis of the barriers to its adoption in the FMCG sector, with a specific focus on Pakistan. By identifying and categorizing these barriers, the chapter offers actionable insights for industry practitioners and policymakers. Furthermore, it emphasizes the need for a strategic approach to overcome these challenges, ensuring that smart warehousing delivers its intended benefits. The findings of this study are particularly relevant for emerging economies, where resource constraints and infrastructural limitations pose additional hurdles to technological adoption.

12.2 Literature Review

12.2.1 Smart Supply Chain: Concept and Design

The concept of the smart supply chain (SSC) has been widely explored, with definitions converging on key attributes. Mubarik and Khan (2024) describe it as a fully integrated system combining advanced technologies such as the Internet of Things (IoT) with contemporary supply chain management (SCM) theories, aiming to create intelligent, automated, and networked processes across organizations. Studies like Li (2023), Li and Huang (2024) emphasize the real-time communication and information symmetry that SSCs enable, spanning logistics, information flow, and capital flow. Mubarik and Khan (2024) highlight the key differentiators of SSCs: effective integration, enhanced visualization, and collaborative capabilities powered by cutting-edge technologies.

Across various definitions, the central characteristics of smart supply chains can be distilled into four core components:

- *Integration: Seamless connectivity among all supply chain nodes using IoT, cloud computing, and advanced analytics.*

- *Automation: Deployment of robotics, autonomous systems, and AI for process optimization.*
- *Visibility: Real-time data sharing and enhanced transparency across supply chain activities.*
- *Intelligence: Leveraging data analytics and AI for predictive decision-making and strategic adaptability.*

For this chapter, SSC is defined as a technology-driven supply chain system that integrates intelligent decision-making, automation, and real-time visibility to enhance efficiency, resilience, and sustainability.

Before we proceed toward as to how smart supply chain has transformed SC performance, we provide a brief recap of evolution of smart supply chain. The evolution of supply chain management to the smart era has been a journey shaped by technological advancements and the growing complexity of global trade. Initially, supply chains relied on manual processes, fragmented communication, and reactive strategies, leading to inefficiencies and high costs. The advent of enterprise resource planning (ERP) systems in the 1980s marked a pivotal shift by enabling integrated data management across functions. The introduction of just-in-time (JIT) manufacturing in the 1990s further emphasized lean inventory practices, pushing warehouses to adopt more streamlined operations. The early 2000s saw the rise of digital technologies, including barcoding and RFID systems, which enhanced tracking and inventory accuracy. However, it was the convergence of Industry 4.0 technologies, such as IoT, AI, and robotics, that truly redefined the supply chain landscape. Smart supply chain management (SSCM) emerged as a holistic framework leveraging these technologies to transform supply chains into dynamic, predictive, and self-optimizing systems (Rahmanzadeh et al. 2023).

Throughout this transformation, warehousing has remained a critical node. Traditional warehouses served as static storage centers, but smart warehousing integrates advanced technologies to enable real-time monitoring, automated inventory management, and dynamic order fulfillment. This evolution positions warehousing as a pivotal enabler of SSCM.

At its core, SSCM is about achieving a synergistic blend of technology and management practices to create agile, resilient, and sustainable supply chains (Zhang et al. 2023; Sadrabadi et al. 2024). By leveraging advanced technologies, SSCM enables organizations to:

- Predict and respond to market changes: Real-time data analytics and predictive algorithms allow for proactive decision-making.
- Enhance collaboration: Cloud platforms and blockchain foster trust and transparency among stakeholders.
- Reduce operational inefficiencies: Automation minimizes manual intervention, reducing errors and costs.

This essence of SSCM lies in its ability to align operational goals with broader strategic objectives, creating value for both organizations and their stakeholders.

12.3 Smart Supply Chain Management and Key Performance Parameters

Smart supply chains significantly contribute to sustainability by addressing environmental, social, and economic dimensions. Advanced technologies within smart supply chains enable precise inventory management and demand forecasting, minimizing waste and reducing overproduction. These efficiencies align with environmental goals by lowering resource consumption and emissions. AI-driven route optimization further reduces fuel usage in logistics, cutting carbon footprints across supply networks. On the social front, automation of hazardous tasks improves workplace safety, while blockchain technologies ensure better compliance tracking, fostering ethical practices (Patidar et al. 2023). Collectively, these advancements create a framework for sustainable development, balancing efficiency with environmental and social responsibility. Resilience is another critical benefit of smart supply chains, as they equip organizations with the tools to anticipate and adapt to disruptions. Advanced analytics and IoT systems provide early warning mechanisms, enabling firms to implement contingency plans effectively. During the COVID-19 pandemic, organizations with smart supply chains demonstrated superior adaptability, maintaining operational continuity and mitigating financial losses. By leveraging predictive insights and fostering flexibility, these systems enhance the overall robustness of supply chains, preparing them to withstand unforeseen challenges and recover swiftly (Lee et al. 2023).

Integration lies at the heart of smart supply chains, fostering seamless connectivity among stakeholders. Interoperable systems unify data from diverse sources, enabling real-time coordination between suppliers, manufacturers, distributors, and retailers. Technologies like IoT and blockchain enhance visibility and trust, creating a collaborative ecosystem that accelerates decision-making and reduces inefficiencies. These integrated networks reduce lead times, optimize resource allocation, and enhance service levels, ensuring that supply chain operations are both synchronized and agile. Smart supply chains also derive strength from their intelligence, driven by big data analytics, AI, and machine learning (Tadayonrad and Ndiaye 2023). These capabilities facilitate predictive and prescriptive decision-making, enabling firms to anticipate market trends, optimize inventory levels, and refine pricing strategies. Predictive maintenance technologies reduce equipment downtime, extending asset lifespans and ensuring cost efficiency. By harnessing intelligent systems, organizations can enhance operational performance and gain strategic insights, empowering them to remain competitive in dynamic markets.

The cumulative impact of smart supply chain management on warehousing is transformative. By integrating sustainability, resilience, integration, and intelligence, smart supply chains elevate warehousing performance to exceptional levels. Real-time inventory visibility powered by IoT and robotics ensures optimal stock levels while minimizing waste. Automated picking systems and AGVs improve order accuracy and speed, reducing labor costs and enhancing customer satisfaction. AI-driven analytics enable precise demand forecasting, optimizing space utilization and lowering storage costs. Real-time monitoring of environmental conditions ensures the safe storage of temperature-sensitive products, further enhancing resilience. Smart systems streamline workflows, eliminate bottlenecks, and drive efficiency across operations.

Ultimately, smart supply chain management transforms warehousing from a static, reactive function into a dynamic, value-adding component of the supply chain. By aligning warehousing with broader strategic goals, it enhances operational performance, ensures long-term sustainability, and positions warehouses as critical enablers of competitive advantage.

12.3.1 Smart Warehousing

Warehousing has long been a vital component of supply chain management, primarily serving as storage hubs for goods awaiting further distribution. Traditionally, these facilities relied heavily on manual processes for operations such as picking, delivery, and inventory management. Warehouse personnel were responsible for the physical handling of goods, often using paper-based systems to track inventory and oversee operations. While indispensable, this traditional approach was riddled with inefficiencies, errors, and delays, making warehouses operationally burdensome and costly (Raman et al. 2023). As a result, warehouses were historically regarded as cost centers—necessary but expensive facilities that added little direct value to the supply chain. The inefficiencies inherent in manual processes and the limited visibility into inventory movement often led to increased labor costs, overstocking, and wastage, further cementing the perception of warehouses as a financial liability.

The advent of modern and smart supply chain technologies has radically transformed warehousing, elevating it from a static, resource-draining function to a dynamic and value-creating operation. Smart warehousing employs advanced technologies such as automation, digitalization, and IoT-driven connectivity to optimize every aspect of warehouse operations. Unlike traditional systems, smart warehouses leverage cyber-physical systems (CPS) to integrate physical processes with digital technologies, enabling real-time monitoring, control, and optimization (Tiwari 2023; Sahara and Aamer 2023). This transition has redefined the role of warehousing, making it an integral part of the smart supply chain ecosystem. Cyber-physical systems facilitate tasks such as efficient data acquisition, accurate inventory repositioning, human activity identification, and multi-robot collaboration, creating a streamlined and highly efficient warehouse environment. These capabilities have turned warehouses into intelligent hubs that enhance operational performance, minimize human intervention, and improve overall supply chain agility.

Definitions of smart warehousing highlight its transformative nature, each focusing on distinct but overlapping aspects. For example, Verma et al. (2023) describe smart warehousing as a CPS-driven model aligned with Industry 4.0 objectives, emphasizing real-time data handling and machine coordination. Various researchers stress the role of IoT technologies, particularly RFID and sensors, in enhancing inventory accuracy and protecting fragile items. Sharma and Tripathi (2024) highlight the role of the integration of autonomous vehicles, such as self-navigating reach trucks, to improve logistical efficiency and adaptability. Other definitions extend to the implementation of fully automated and paperless systems, with some designs including AI-powered predictive analytics for proactive decision-making. Despite the variation in emphasis, all definitions converge on certain key characteristics: the integration of automation, digitalization, and connectivity; real-time data analytics; and the minimization of human intervention in operational workflows.

From these definitions, a unified concept emerges: smart warehousing is an advanced warehouse management system that integrates cyber-physical systems, IoT technologies, and automation to create a dynamic, data-driven, and self-optimizing operational environment. This definition underscores the centrality of real-time connectivity, decision-making capabilities, and enhanced efficiency, differentiating smart warehousing from its traditional counterparts. Smart warehouses are no longer static storage spaces but intelligent nodes in the supply chain that actively contribute to its resilience, sustainability, and adaptability. As these technologies evolve, so too does the design of smart warehouses. RFID and sensor-based tracking systems provide unparalleled accuracy in real-time object identification, ensuring the efficient handling of sensitive and fragile goods. Autonomous vehicles within warehouses navigate dynamically, adapting to environmental changes and eliminating the inefficiencies of pre-programmed paths. These innovations represent a progression from basic automation to intelligent environments capable of autonomous decision-making. Smart warehousing has redefined the role of storage facilities, transforming them into strategic assets that enhance supply chain performance, responsiveness, and competitiveness.

12.4 Challenges in Smart Warehouse Adoption

The adoption of smart warehousing has garnered considerable attention in recent years due to its potential to revolutionize supply chain operations. However, despite its transformative advantages, the implementation of smart warehousing systems faces several significant challenges. These challenges are rooted in both technological and organizational factors, as well as external constraints such as economic conditions and regulatory frameworks. A review of the literature reveals a diverse set of barriers that organizations encounter when transitioning to smart warehousing, which often hampers their ability to fully realize its benefits (Plakantara et al. 2024).

One of the most cited challenges is the lack of infrastructure. Smart warehousing requires a robust digital and physical infrastructure, including high-speed Internet connectivity, IoT devices, and automated equipment. Many organizations, particularly in developing economies, lack the foundational infrastructure necessary to support these advanced technologies. This deficiency not only delays adoption but also raises the overall cost of implementation. Closely related is the challenge of huge capital investment costs. The initial setup of a smart warehouse involves substantial expenditure on advanced technologies such as robotics, sensors, and AI-driven software. For many companies, especially small and medium-sized enterprises (SMEs), these costs are prohibitively high, discouraging investment despite the long-term operational savings smart warehousing can offer. Technical complexities also pose a significant hurdle. Smart warehousing relies on the seamless integration of various technologies, such as cyber-physical systems, IoT, and AI. Ensuring that these technologies communicate effectively requires highly specialized technical expertise, which is often scarce. Inadequate training for existing staff further exacerbates the issue, leading to operational inefficiencies during and after implementation. The complexity of integrating smart warehousing systems with existing enterprise resource planning (ERP) systems and legacy infrastructure adds another layer of difficulty, often resulting in data inconsistencies and workflow disruptions (Mostafa et al. 2018).

Data management emerges as another critical challenge. Smart warehouses generate and process enormous volumes of data in real time, ranging from inventory levels to predictive analytics for demand forecasting. Managing this data effectively requires robust data storage and processing capabilities, as well as advanced analytics tools. Organizations often struggle to ensure data quality, consistency, and security, particularly in the context of increasing cyber threats. The sheer volume of data generated can overwhelm traditional data management systems, creating bottlenecks that undermine the efficiency gains smart warehousing promises. The challenge of workforce adaptation is also widely discussed in the literature (Liu et al. 2022). Transitioning to smart warehousing often leads to resistance from employees, who may perceive automation as a threat to job security. Additionally, the need for specialized skills to operate and maintain smart warehousing technologies creates a skill gap that organizations must address through training and upskilling initiatives. Failure to do so can result in operational disruptions and underutilization of the smart warehousing system's capabilities (Kuo et al. 2021).

Cybersecurity concerns further complicate the adoption of smart warehousing. As these systems rely heavily on IoT devices and cloud-based platforms, they are vulnerable to cyberattacks. Breaches can compromise sensitive data and disrupt warehouse operations, causing significant financial and reputational damage. Organizations must invest in robust cybersecurity measures, adding to the overall cost and complexity of adoption. Economic and market factors also play a significant role in impeding the adoption of smart warehousing. Economic downturns and uncertainty can lead organizations to deprioritize capital-intensive projects like smart warehousing. Additionally, fluctuations in market demand can make it challenging to justify the investment, particularly when the return on investment (ROI) is not immediately apparent. Regulatory barriers, such as stringent data protection laws and industry-specific compliance requirements, further complicate the process, as organizations must ensure that their smart warehousing systems meet all legal standards. Organizational inertia is another critical barrier. Many organizations are reluctant to change long-established processes, particularly if their current systems are perceived

Table 12.1 Challenges of smart warehousing adoption (Priority value)

Lack of infrastructure	Regulatory barriers
Huge capital investment cost	Organizational inertia
Technical complexities	Lack of top-management support
Data management	Geographical constraints
A huge volume of data	Environmental sustainability concerns
Workforce resistance	Cultural resistance to innovation
Skill gaps	Lack of awareness about benefits
Cybersecurity concerns	Insufficient training programs
Integration with legacy systems	High maintenance costs
Data quality and consistency	Supply chain partner alignment
Economic uncertainty	Long ROI timelines
Market demand fluctuations	

as adequate. This resistance to change often stems from a lack of awareness or understanding of the benefits of smart warehousing, as well as concerns about the risks and costs involved. A lack of top-management support can further stymie adoption, as strategic decisions to invest in smart warehousing often depend on executive buy-in.

Environmental factors also influence the adoption of smart warehousing. For instance, geographical constraints, such as remote locations or poor connectivity in rural areas, can limit the feasibility of implementing IoT and cloud-based technologies. Furthermore, the need for sustainable practices in warehousing adds another layer of complexity, as organizations must balance the adoption of smart technologies with their environmental impact. Finally, cultural factors, such as varying attitudes toward technology adoption across regions and industries, can influence the speed and extent of smart warehousing implementation. For example, industries with a traditionally conservative approach to innovation may be slower to adopt smart warehousing technologies compared to more tech-savvy sectors. This cultural reluctance can create disparities in adoption rates, even within the same supply chain ecosystem (Table 12.1).

Following tables provide key challenges in the light of literature:

12.5 Method

The mean value approach introduced by Mubarik (2016) and later refined by Mubarik et al. (2018) provides a systematic and quantitative method for prioritizing challenges or factors based on their perceived importance among respondents. This approach employs a three-point Likert scale questionnaire, where respondents rate each challenge or factor as either *Not Important (NI), Somewhat Important, or Important*. These responses are numerically coded, assigning a score of 1 to *Not Important,* 2 to *Somewhat Important*, and 3 to *Important*. This scoring system forms the basis for calculating a mean priority score, which reflects the relative importance of each challenge or factor. To determine the mean value for each challenge, the responses are analyzed by calculating the percentage of respondents selecting each category. For instance, if 20% of respondents rated a challenge as not important, 50% as somewhat important, and 30% as important, these percentages are multiplied by their respective numerical values. In this example, the calculation would involve multiplying 20% by 1, 50% by 2, and 30% by 3. The resulting values are summed to generate a composite mean priority score for that challenge. This score provides a numerical representation of its overall perceived importance, with higher scores indicating greater relevance and priority. Once the mean priority scores for all challenges have been calculated, a cut-off value is determined to differentiate between challenges deemed relevant and those considered less significant. This cut-off value is obtained by averaging the minimum and maximum mean priority scores within the dataset. Challenges with mean priority scores exceeding this cut-off are classified as important and relevant, while those below the cut-off are considered less critical. This threshold ensures a clear and objective basis for prioritization, enabling researchers and practitioners to focus their efforts on the most pressing challenges.

The strength of this approach lies in its ability to capture a nuanced understanding of respondents' perspectives. By incorporating the full distribution of responses rather than relying on simple majority or median scores, the method ensures that the prioritization process reflects the collective judgment of the respondent group. Moreover, the use of a weighted scoring system accommodates varying degrees of importance, providing a more detailed and accurate assessment of each challenge's significance. The mean value approach has been particularly effective in research settings where the objective is to identify and prioritize challenges across diverse stakeholder groups. The methodology's reliance on straightforward calculations and a clear threshold makes it accessible and replicable in various research and practical contexts.

12.6 Context

The fast-moving consumer goods (FMCG) sector in Pakistan plays a pivotal role in the country's economy, contributing significantly to its industrial output, employment generation, and consumer market dynamics. This sector encompasses a wide range of products, including food and beverages, personal care items, household goods, and over-the-counter medicines, characterized by high turnover rates and frequent consumer purchases. Pakistan's FMCG industry has shown consistent growth over the years, driven by a rising population, urbanization, and increasing disposable income among the middle class. With a population exceeding 240 million, a median age of 23 years, and an expanding urban demographic, Pakistan presents a vibrant consumer base that propels the demand for FMCG products (Malik 2019; Javed 2022). The FMCG supply chain in Pakistan is intricate and operates in a challenging environment, making it an ideal context for studying smart warehousing adoption. This sector faces unique logistical complexities due to the high volume and diversity of products, short product lifecycles, and the need for rapid replenishment. These challenges are further exacerbated by infrastructural limitations, fragmented supply chains, and inconsistent regulatory enforcement. For instance, perishable goods like dairy and fresh produce require stringent temperature controls, while non-perishable items demand precise inventory management to avoid overstocking or pilferage. These factors underscore the critical role of warehousing in ensuring operational efficiency and product availability in the FMCG sector. Pakistan's FMCG sector operates in a dual-channel system, serving both modern retail outlets and traditional markets. The latter dominates the market landscape, accounting for approximately 70% of consumer goods distribution. This duality presents additional challenges in supply chain management, as traditional markets often rely on fragmented logistics networks and manual warehousing practices.

12.7 Findings

12.7.1 Profile of Participants

The expert profile for the study comprises 49 participants, showcasing a diverse mix of professional experience and hierarchical designations, reflecting a comprehensive representation of the FMCG sector. Regarding experience, the majority of experts, 67% (33 individuals), possess more than five years of experience in their respective fields, indicating substantial practical knowledge. Furthermore, 29% (14 participants) bring over ten years of expertise, while a smaller group, 4% (2 individuals), has over two decades of professional experience, providing deep industry insights. From a designation perspective, the sample includes a range of managerial and operational roles. Managers and officers/executives each constitute 22% (11 participants) of the sample, representing a balance between decision-making and ground-level perspectives. Deputy managers account for 18% (9 individuals), followed by Assistant Managers at 16% (8 individuals), ensuring mid-level management representation. Senior-level employees contribute 12% (6 individuals), while general managers and directors each form 4% (2 participants), adding strategic viewpoints to the dataset. This diverse mix of experience and roles ensures a holistic understanding of the challenges and intricacies associated with smart warehousing adoption, offering both strategic and operational insights (Table 12.2).

Table 12.2 Profile of respondents

Category	N (49)	(%)
Experience (n 49)		
5 + years experience	33	67
10 + years experience	14	29
20 + years experience	2	4
Designation (n 49)		
Manager	11	22
Deputy manager	9	18
Assistant manager	8	16
Senior-level employee	6	12
General manager	2	4
Director	2	4
Officers/Exec	11	22

12.7.2 Findings

The findings of this study highlight the key challenges associated with the adoption of smart warehousing technologies, as perceived by industry experts in the FMCG sector of Pakistan. Using the mean priority value (MPV) approach, the study categorizes these challenges based on their importance, with a calculated cut-off value of 1.695. Barriers exceeding this value are considered high-priority challenges. These findings provide critical insights into the complexities of transitioning from traditional to smart warehousing in a developing market context, aligning with and extending existing literature on the subject. The results indicate that the most significant barriers to smart warehousing adoption include economic uncertainty, integration with legacy systems, skill gaps, workforce resistance, lack of infrastructure, huge capital investment costs, and technical complexities. Each of these challenges reflects a combination of technological, organizational, and market-driven factors that impede the adoption process. Economic uncertainty, with the highest MPV (2.88), underscores the financial volatility and risk aversion prevalent in the FMCG sector, particularly in developing economies. This finding corroborates prior research, such as that by Nasiri et al., which highlights the impact of macroeconomic conditions on firms' willingness to invest in capital-intensive technologies like smart warehousing.

Integration with legacy systems (MPV 2.87) emerges as another critical challenge, illustrating the difficulties organizations face in aligning outdated infrastructure with advanced technologies. This is consistent with Hofmann and Rüsch, who emphasize the complexities of synchronizing IoT devices and automated systems with pre-existing workflows and ERP platforms. Such misalignments often result in data inconsistencies, operational inefficiencies, and escalated costs, which deter firms from pursuing smart warehousing solutions. Skill gaps (MPV 2.78) and workforce resistance (MPV 2.63) represent significant organizational barriers, reflecting the human element of technology adoption. The transition to smart warehousing requires a workforce proficient in handling advanced technologies such as robotics, AI, and IoT. However, the lack of specialized training programs and resistance to change among employees impede this shift. These findings align with Ralston and Blackhurst, who note that workforce adaptability and skill development are crucial for successful Industry 4.0 transformations. Additionally, workforce resistance echoes organizational inertia, as identified in the literature, where fear of job displacement and unfamiliarity with new systems create reluctance to embrace automation. The lack of infrastructure (MPV 2.63) and the high costs associated with capital investment (MPV 2.61) are among the most pressing barriers, particularly in the context of developing economies. Pakistan's underdeveloped infrastructure, including unreliable Internet connectivity and inadequate warehousing facilities, poses significant challenges to the implementation of smart technologies. These findings resonate with studies by Liu et al. (2022), which stress the need for robust foundational infrastructure to support IoT and cyber-physical systems. The high capital costs further exacerbate this issue, as firms struggle to justify such expenditures without immediate and tangible returns. Technical complexities (MPV 2.59) and data management challenges (MPV 2.43) also rank high among the barriers. The former underscores the difficulties in integrating multiple technologies, ensuring interoperability, and managing complex workflows. The latter highlights the challenges of handling large volumes of data generated by smart warehouses, including ensuring data quality, consistency, and security. These findings are supported by Chen et al., who emphasize that the technical sophistication of smart systems demands advanced expertise and robust data governance frameworks to avoid inefficiencies and security risks.

The study also identifies several barriers with lower MPVs, such as cultural resistance to innovation (1.1), regulatory barriers (1.13), and environmental sustainability concerns (1.05). While these barriers are less critical than the high-priority challenges, they still highlight underlying issues that may hinder adoption in specific contexts. For instance, cultural resistance and organizational inertia reflect deep-seated attitudes toward change and innovation, consistent with the findings of Chen et al. Similarly, regulatory barriers, though less prominent in this study, have been highlighted in global literature as significant obstacles in heavily regulated industries.

In the light of existing literature, the findings of the study align with broader themes while offering unique insights specific to the FMCG sector in a developing economy. Prior research has extensively discussed the role of infrastructure, cost, and technical challenges in smart warehousing adoption, but this study emphasizes how these factors interact with market-specific dynamics such as economic volatility and workforce resistance. Additionally, the emphasis on data management and cybersecurity challenges reflects the increasing digitalization of supply chain operations and the associated risks (Table 12.3).

High-priority barriers also include cybersecurity concerns (MPV 2.01), insufficient training programs (MPV 1.99), and long return-on-investment (ROI) timelines (MPV 2.11). Cybersecurity concerns reflect the vulnerabilities associated with IoT devices and cloud-based systems, as highlighted by Ghobakhloo and Fathi. Insufficient training programs indicate a gap in organizational readiness, underscoring the

Table 12.3 Challenges of smart warehousing adoption

Challenges	MPV
Supply chain partner alignment	0.51
Environmental sustainability concerns	1.05
Organizational inertia	1.05
Cultural resistance to innovation	1.1
Regulatory barriers	1.13
A huge volume of data	1.15
Lack of top-management support	1.21
Market demand fluctuations	1.25
Lack of awareness about benefits	1.28
Data quality and consistency	1.32
Geographical constraints	1.54
High maintenance costs	1.86
Insufficient training programs	1.99
Cybersecurity concerns	2.01
Long ROI timelines	2.11
Data management	2.43
Technical complexities	2.59
Huge capital investment cost	2.61
Lack of infrastructure	2.63
Workforce resistance	2.63
Skill gaps	2.78
Integration with legacy systems	2.87
Economic uncertainty	2.88

*Mean Priority value (Cut-off value 1.695)

need for structured initiatives to upskill employees and familiarize them with new technologies. Long ROI timelines further deter adoption, as organizations prioritize short-term financial stability over long-term strategic gains, a trend noted in existing research on technology adoption in cost-sensitive industries (Table 12.4).

12.8 Conclusion and Managerial Implications

This chapter set out to examine the challenges of adopting smart warehousing technologies within the fast-moving consumer goods (FMCG) sector, focusing on identifying and prioritizing barriers that hinder this transition. By leveraging expert insights, the study aimed to provide a deeper understanding of the factors

Table 12.4 High-priority barriers

High maintenance costs
Insufficient training programs
Cybersecurity concerns
Long ROI timelines
Data management
Technical complexities
Huge capital investment cost
Lack of infrastructure
Workforce resistance
Skill gaps
Integration with legacy systems
Economic uncertainty

affecting smart warehousing adoption and offer actionable implications for organizations navigating this complex process. The context of this research was Pakistan's FMCG sector, a dynamic industry marked by high product diversity, rapid replenishment needs, and an evolving competitive landscape. This context provided a unique lens to explore the challenges in a developing economy, where infrastructural constraints and economic uncertainties amplify the difficulties associated with adopting advanced supply chain technologies. Data for the study were collected from 49 experts in the FMCG sector, spanning a wide range of roles and experience levels. The respondents, including managers, directors, senior employees, and operational executives, brought diverse perspectives, allowing for a holistic view of the challenges. Using a three-point Likert scale questionnaire, participants rated various barriers to smart warehousing adoption based on their perceived importance. The mean priority value (MPV) approach was employed to quantify these perceptions, combining the percentage of respondents for each rating with a weighted score. This method provided a rigorous and transparent means of prioritizing challenges, with a calculated cut-off value of 1.695 distinguishing high-priority barriers from less critical ones. The findings revealed several key barriers to smart warehousing adoption, with economic uncertainty, integration with legacy systems, skill gaps, workforce resistance, lack of infrastructure, and high capital investment costs ranking as the most significant challenges. The results underscore the interplay of technological, organizational, and market-specific factors that complicate the transition to smart warehousing. Notably, economic uncertainty emerged as the highest-priority barrier, highlighting the impact of financial instability and risk aversion on firms' willingness to invest in advanced technologies. This finding is particularly relevant in the context of Pakistan, where macroeconomic volatility and inflationary pressures exacerbate financial constraints for businesses (Fig. 12.1).

Another critical finding was the challenge of integrating smart technologies with existing legacy systems. This issue reflects the technical complexities of ensuring interoperability between new and old systems, a problem that often leads to data

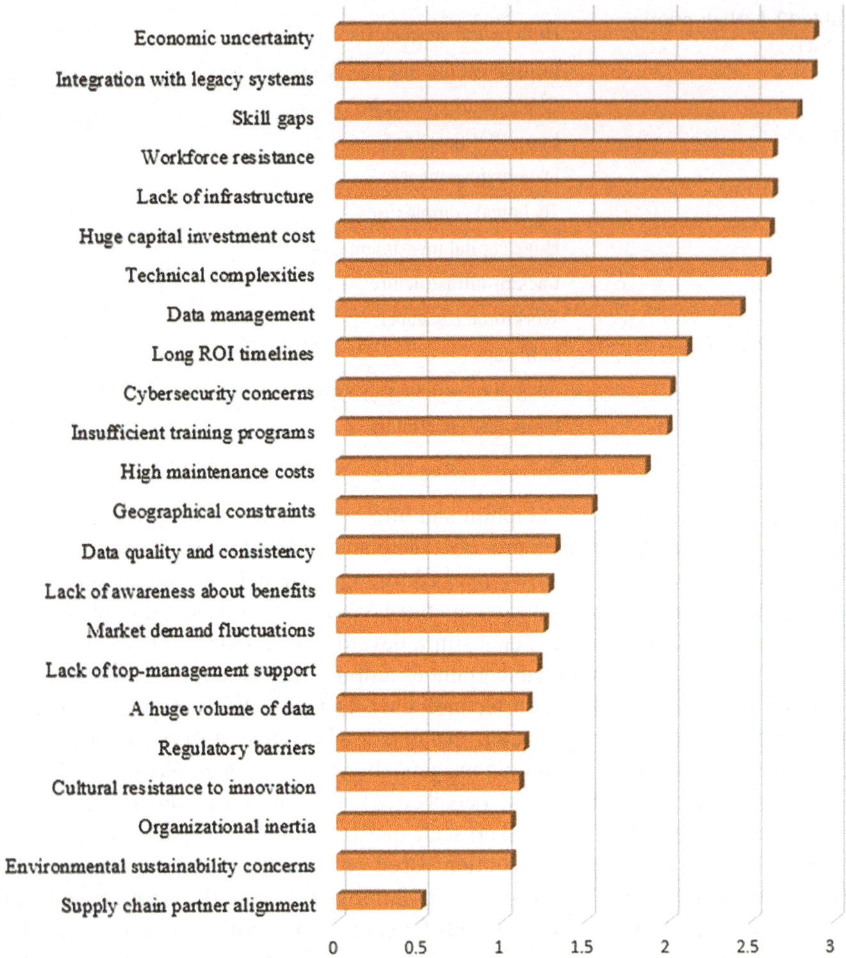

Fig. 12.1 MPVs of barriers

inconsistencies and operational inefficiencies. The importance of addressing this barrier is echoed in global literature, but the study's focus on Pakistan's FMCG sector highlights how infrastructural limitations and fragmented supply chains intensify these difficulties. Similarly, the findings on skill gaps and workforce resistance emphasize the human element of technology adoption, revealing that the lack of specialized expertise and resistance to change among employees are significant hurdles. These barriers underscore the need for targeted training programs and cultural shifts within organizations to facilitate the adoption process.

The findings of this study contribute uniquely to the literature on smart supply chains by providing a granular analysis of the challenges specific to the FMCG sector in a developing economy. While much of the existing research focuses on the technological capabilities and benefits of smart warehousing, this chapter sheds light on the barriers that impede adoption, particularly in resource-constrained environments. By identifying high-priority challenges such as insufficient training programs, cybersecurity concerns, and long ROI timelines, the study expands the scope of existing knowledge and provides a more comprehensive understanding of the adoption landscape. This contribution is particularly significant given the increasing emphasis on smart supply chain technologies as a means of achieving operational efficiency, sustainability, and resilience. For firms interested in adopting smart warehousing, the findings have several practical implications. First, organizations must address the issue of workforce readiness by investing in upskilling initiatives and creating a culture that embraces technological innovation. Training programs should focus on developing the technical expertise required to operate and maintain smart warehousing systems, as well as fostering an understanding of their strategic value. This approach can help mitigate resistance to change and ensure that employees are equipped to leverage the full potential of advanced technologies. Second, firms need to prioritize infrastructure development and integration capabilities. This includes upgrading legacy systems to enable seamless communication with new technologies and investing in robust digital infrastructure to support IoT devices, sensors, and real-time data analytics. Collaboration with supply chain partners can also play a critical role in addressing integration challenges, as it fosters alignment and consistency across the supply chain network. By taking a proactive approach to infrastructure and integration, firms can reduce operational bottlenecks and enhance the effectiveness of smart warehousing solutions.

Third, organizations must carefully evaluate the financial feasibility of adopting smart warehousing technologies. While the high capital investment costs and long ROI timelines are significant barriers, firms can overcome these challenges by adopting phased implementation strategies or seeking external funding through partnerships, government incentives, or grants. Additionally, conducting comprehensive cost–benefit analyses can help firms identify areas where smart warehousing can deliver the most value, enabling targeted investments that maximize returns. Cybersecurity is another critical area that firms must address when transitioning to smart warehousing. The reliance on IoT and cloud-based systems makes smart warehouses vulnerable to cyberattacks, which can compromise sensitive data and disrupt operations. Implementing robust cybersecurity measures, such as encryption, firewalls, and regular system audits, is essential to protect against these risks. Firms should also prioritize data management practices to ensure the quality, consistency, and security of the vast volumes of data generated by smart warehousing systems.

Finally, the findings highlight the importance of leadership commitment in driving smart warehousing adoption. Top-management support is crucial for securing the resources and organizational alignment needed to overcome adoption barriers. Leaders must champion the adoption process, communicate its strategic value, and create an environment that fosters innovation and collaboration. By doing so, they can build momentum for change and ensure that smart warehousing initiatives are successfully implemented.

References

Allison, J. E. (2023). Unsustainable: Amazon, warehousing, and the politics of exploitation, Univ of California Press.

Ali, S. S., Kaur, R., Gupta, H., Ahmad, Z., & Jebahi, K. (2024). A decision-making framework for determinants of an organisation's readiness for smart warehouse. *Production Planning & Control*, *35*(14), 1887-1908.

Dantas, R., Sabir, I., Martins, J. M., Majid, M. B., Rafiq, M., Martins, J. N., & Rana, K. (2023). Role of green and multisensory packaging in environmental sustainability: Evidence from FMCG sector of Pakistan. *Cogent Business & Management*, *10*(3), 2285263.

Javed, H. (2022, July 2). Pakistan. Retrieved from Startup Pakistan: https://startuppakistan.com.pk/top-fmcg-companies-in-pakistan/

Kuo, T. C., Chen, K. J., Shiang, W. J., Huang, P. T. B., Otieno, W., & Chiu, M. C. (2021). Collaborative data-driven analytics of material resource management in the smart supply chain by using a hybrid Industry 3.5 strategy. Resources, Conservation, and Recycling, p. 164. https://doi.org/10.1016/j.resconrec.2020.105160

Kukard, W. A. (2023). Transforming an Industry: Leveraging Emerging Technologies in the Smart Home Environment to Create Value for the FMCG Industry.

Li, Y., & Huang, H. (2024). Efficient Task Planning for Heterogeneous AGVs in Warehouses. *IEEE Transactions on Intelligent Transportation Systems*.

Lee, K. L., Wong, S. Y., Alzoubi, H. M., Al Kurdi, B., Alshurideh, M. T., & El Khatib, M. (2023). Adopting smart supply chain and smart technologies to improve operational performance in manufacturing industry. *International Journal of Engineering Business Management*, *15*, 18479790231200614.

Li, K. (2023). Optimizing warehouse logistics scheduling strategy using soft computing and advanced machine learning techniques. *Soft Computing*, *27*(23), 18077-18092.

Liu, C., Ji, H., & Wei, J. (2022). Smart Supply Chain Risk Assessment in Intelligent Manufacturing. Journal of Computer Information Systems, 62(3), 609–621. https://doi.org/10.1080/08874417.2021.1872045

Mahdzir, N., Rou, S. J., Zeng, L. F., Din, M. A., Suliman, K. R., & Bakar, M. A. (2023, February). The Consequences of COVID-19 Pandemic on FMCG Warehouse Operations in Klang Valley: A Qualitative Study Approach. In *Proceedings of the 10th International Conference on Business, Accounting, Finance and Economics (BAFE 2022)* (Vol. 234, p. 239). Springer Nature.

Malik, W. (2019). The FMCG Market and The Pakistani Economy. Retrieved from Crux: https://www.crux.pk/business/the-fmcg-market-and-the-pakistani-economy/

Mostafa, N. A., Hamdy, W., Mostafa, N., & Elawady, H. (2018). Towards a Smart Warehouse Management System History of Science View project Towards a Smart Warehouse Management System. https://www.researchgate.net/publication/328007244

Mubarik, M. S., & Khan, S. A. (2024). *The Theory, Methods and Application of Managing Digital Supply Chains*. Emerald Publishing Limited.

Mubarik, M. S. (2016). *Human capital and performance of small and medium manufacturing enterprises: A study of Pakistan* (Doctoral dissertation, Jabatan Ekonomi, Fakulti Ekonomi dan Pentadbiran, Universiti Malaya).

Mubarik, M. S., Chandran, V. G. R., & Devadason, E. S. (2018). Measuring human capital in small and medium manufacturing enterprises: what matters?. *Social indicators research*, *137*, 605-623.

Patidar, A., Sharma, M., Agrawal, R., & Sangwan, K. S. (2023). Supply chain resilience and its key performance indicators: an evaluation under Industry 4.0 and sustainability perspective. *Management of Environmental Quality: An International Journal*, *34*(4), 962–980.

Plakantara, S. P., Karakitsiou, A., & Mantzou, T. (2024). Managing Risks in Smart Warehouses from the Perspective of Industry 4.0. In *Disruptive Technologies and Optimization Towards Industry 4.0 Logistics* (pp. 1–47). Cham: Springer International Publishing.

Raman, R., Ramalingam, L., Sutaria, K. K., Kamthan, M., Sangeetha, S., & Murugan, S. (2023, November). Smart Warehouse Solutions for Efficient Onion Buffer Stock Management System. In *2023 7th International Conference on Electronics, Communication and Aerospace Technology (ICECA)* (pp. 1260–1265). IEEE.

Rahmanzadeh, S., Pishvaee, M. S., & Govindan, K. (2023). Emergence of open supply chain management: the role of open innovation in the future smart industry using digital twin network. *Annals of Operations Research, 329*(1), 979-1007.

Sadrabadi, M. H. D., Makui, A., Ghousi, R., & Jabbarzadeh, A. (2024). Designing a green-resilient closed-loop supply chain to preserve business continuity under interrelated disruptions (a case study of the automotive battery industry). *Computers & Chemical Engineering, 185*, 108675.

Sahara, C. R., & Aamer, A. M. (2023, April). Real-time data integration in smart warehouse as a contemporary approach. In *AIP Conference Proceedings* (Vol. 2646, No. 1). AIP Publishing.

Sharma, D., & Tripathi, K. (2024). Navigating the Digital Age: Smart Manufacturing and Logistics. In *Recent Trends in Artificial Intelligence Towards a Smart World: Applications in Industries and Sectors* (pp. 99–144). Singapore: Springer Nature Singapore.

Tiwari, S. (2023). Smart warehouse: A bibliometric analysis and future research direction. *Sustainable Manufacturing and Service Economics, 2*, 100014.

Tadayonrad, Y., & Ndiaye, A. B. (2023). A new key performance indicator model for demand forecasting in inventory management considering supply chain reliability and seasonality. *Supply Chain Analytics, 3*, 100026.

Verma, A., Tripathy, S., & Singhal, D. (2023). The significance of warehouse management in supply chain: An ISM approach. *Decision Making: Applications in Management and Engineering, 6*(1), 92-110.

Zhang, G., Yang, Y., & Yang, G. (2023). Smart supply chain management in Industry 4.0: the review, research agenda and strategies in North America. *Annals of operations research, 322*(2), 1075–1117.

Zhen, L., & Li, H. (2022). A literature review of smart warehouse operations management. *Frontiers of Engineering Management, 9*(1), 31-55.

The manufacturer's authorised representative in the EU is Springer Nature Customer Service Centre GmbH, Europaplatz 3, 69115 Heidelberg, Germany. If you have any concerns regarding our products, please contact ProductSafety@springernature.com

Printed and bound by CPI Group (UK) Ltd, Croydon, CR0 4YY
30/09/2026
02225300-0008